Ben Stacy Jerrik (Ed.)

Kulířov

Ben Stacy Jerrik (Ed.)

Kulířov

Karviná District, Village, Okres, Moravian–Silesian Region

Part Press

Contents

Articles

References

Kulířov

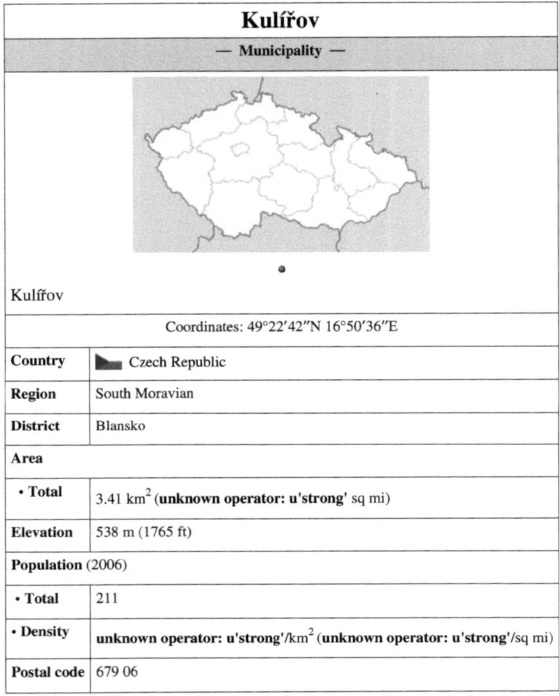

<table>
<tr><td colspan="2" align="center">Kulířov</td></tr>
<tr><td colspan="2" align="center">— Municipality —</td></tr>
<tr><td colspan="2" align="center">Kulířov</td></tr>
<tr><td colspan="2" align="center">Coordinates: 49°22′42″N 16°50′36″E</td></tr>
<tr><td>Country</td><td>Czech Republic</td></tr>
<tr><td>Region</td><td>South Moravian</td></tr>
<tr><td>District</td><td>Blansko</td></tr>
<tr><td>Area</td><td></td></tr>
<tr><td>• Total</td><td>3.41 km^2 (unknown operator: u'strong' sq mi)</td></tr>
<tr><td>Elevation</td><td>538 m (1765 ft)</td></tr>
<tr><td>Population (2006)</td><td></td></tr>
<tr><td>• Total</td><td>211</td></tr>
<tr><td>• Density</td><td>unknown operator: u'strong'/km^2 (unknown operator: u'strong'/sq mi)</td></tr>
<tr><td>Postal code</td><td>679 06</td></tr>
</table>

Kulířov is a village and municipality (*obec*) in Blansko District in the South Moravian Region of the Czech Republic.

The municipality covers an area of 3.41 square kilometres (**unknown operator: u'strong'** sq mi), and has a population of 211 (as at 3 July 2006).

Kulířov lies approximately 15 kilometres (9 mi) east of Blansko, 26 km (16 mi) north-east of Brno, and 192 km (119 mi) south-east of Prague.

References

- Czech Statistical Office: Municipalities of Blansko District [1]

References

[1] http://www.czso.cz/lexikon/mos_vdb.nsf/okresy/CZ0621/

Village

A **village** is a clustered human settlement or community, larger than a hamlet with the population ranging from a few hundred to a few thousand (sometimes tens of thousands). Though often located in rural areas, the term urban village is also applied to certain urban neighbourhoods, such as the East Village in Manhattan, New York City and the Saifi Village in Beirut, Lebanon, as well as Hampstead Village in the London conurbation. Villages are normally permanent, with fixed dwellings; however, transient villages can occur. Further, the dwellings of a village are fairly close to one another, not scattered broadly over the landscape, as a dispersed settlement.

The village of Rougon in Provence, France

In the past, villages were a usual form of community for societies that practise subsistence agriculture, and also for some non-agricultural societies. In Great Britain, a hamlet earned the right to be called a village when it built a church.[1] In many cultures, towns and cities were few, with only a small proportion of the population living in them. The Industrial Revolution attracted people in larger numbers to work in mills and factories; the concentration of people caused many villages to grow into towns and cities. This also enabled specialization of labor and crafts, and development of many trades. The trend of urbanization continues, though not always in connection with industrialisation. Villages have been eclipsed in importance as units of human society and settlement.

An alpine village in the Lötschental Valley, Switzerland

Traditional villages

Although many patterns of village life have existed, the typical village was small, consisting of perhaps 5 to 30 families. Homes were situated together for sociability and defence, and land surrounding the living quarters was farmed. Traditional fishing villages were based on artisan fishing and located adjacent to fishing grounds.

Berber village in Ourika valley, High Atlas, Morocco.

South Asia

India "The soul of India lives in its villages", declared M. K. Gandhi[2] at the beginning of 20th century. According to the 2011 census of India, 68.84% of Indians (around 833.1 million people) live in 640,867 different villages.[3] The size of these villages varies considerably. 236,004 Indian villages have a population less than 500, while 3,976 villages have a population of 10,000+. Most of the villages have their own temple, mosque or church depending on the local religious following.

A North Indian village in Rajasthan, India

A South Indian Village in Andhra Pradesh, India

East Asia

People's Republic of China In mainland China, villages are divisions under township Zh:or town Zh:.

Republic of China (Taiwan) In the Republic of China (Taiwan), villages are divisions under townships or county-controlled cities. The village is called a *tsuen* or *cūn* () under a rural township () and a *li* () under an urban township () or a county-controlled city. See also Li (unit).

Japan

South Korea

Shirakawa-gō, Gifu Japan.

Southeast Asia

Thailand

Brunei, Indonesia, Malaysia and Singapore

In Indonesia, depending on the principles they are administered, villages are called *desa* or *kelurahan*. A *desa* (a term that derives from a Sanskrit word meaning "country" that is found in a name such as "Bangladesh") is administered according to traditions and customary law (*adat*), while a *kelurahan* is administered along more "modern" principles. *Desa* are generally located in rural areas while *kelurahan* are generally urban subdivisions. A village head is respectively called *kepala desa* or *lurah*. Both are elected by the local community. A *desa* or *kelurahan* is itself the subdivision of a *kecamatan* (district), in turn the subdivision of a *kabupaten* (regency).

The *nagari* of Pariangan, West Sumatra.

The same general concept applies all over Indonesia. How ever, there is some variation among the vast numbers of Austronesian ethnic groups. For instance, in Bali villages have been created by grouping traditional hamlets or *banjar*, which constitute the basis of Balinese social life. In the Minangkabau country in West Sumatra province traditional villages are called *nagari* (a term deriving from another Sanskrit word meaning "city", which can be found in a name like "Srinagar"). In some areas such as Tanah Toraja, elders take turns watching over the village at a command post. As a general rule, *desa* and *kelurahan* are groupings of hamlets (*kampung* in Indonesian, *dusun* in the Javanese language, *banjar* in Bali).

In Malaysia, the term *kampung* (sometimes spelling *kampong*) in the English language has been defined specifically as "a Malay hamlet or village in a Malay-speaking country".[4] In other words, a **kampung** is defined today as a village in Brunei, Indonesia, Singapore, and Malaysia. In Malaysia, a *kampung* is determined as a locality with 10,000 or fewer people. Since historical times, every Malay village came under the leadership of a *penghulu* (village chief), who has the power to hear civil matters in his village (see Courts of Malaysia for more details). A Malay village typically contains a *"masjid"* (mosque) or *"surau"* (Muslim chapel), paddy fields and Malay houses on stilts. Malay and Indonesian villagers practice the culture of helping one another as a community, which is better known as "joint bearing of burdens" (*gotong royong*),[5] as well as being family-oriented (especially the concept of respecting one's family [particularly the parents and elders]), courtesy and believing in God (*"Tuhan"*) as paramount to everything else. It is common to see a cemetery near the mosque, as all Muslims in the Malay or Indonesian village want to be prayed for, and to receive Allah's blessings in the afterlife. While in Sarawak and East Kalimantan, some villages are called 'long', primarily inhabited by the Orang Ulu.

Singapore also follows the Malaysian *kampung*. However, there are only a few *kampung* villages remaining, mostly on islands surrounding Singapore such as Pulau Ubin. In the past, there was many *kampung* villages in Singapore but now there aren't many on the mainland.

The term "kampung", sometimes spelled "kampong" is one of many Malay words to have entered common usage in Malaysia and Singapore. Locally, the term is frequently used to refer to one's hometown.

Philippines

In urban areas of the Philippines, the term "village" most commonly refers to private subdivisions, especially gated communities. These villages emerged in the mid-20th century and were initially the domain of elite urban dwellers. Those are common in major cities in the country and their residents have a wide range of income levels. Such villages may or may not correspond to administrative units (usually barangays) and/or be privately administered. Barangays more correspond to the villages of old times, and the chairman (formerly a village datu) now settles administrative, intrapersonal, and political matters or polices the village, though with much less authority and respect than in Indonesia or Malaysia.

Vietnam

Village, or "làng", is a basis of Vietnam society. Vietnam's village is the typical symbol of Asian agricultural production. Vietnam's village typically contains: a village gate, "lũy tre" (bamboo hedges), "đình làng" (communal house) where "thành hoàng" (tutelary god) is worshiped, a common well, "đồng lúa" (rice field), "chùa" (temple) and houses of all families in the village. All the people in Vietnam's villages usually have a blood relationship. They are farmers who grow rice and have the same traditional handicraft. Vietnam's villages have an important role in society (Vietnamese saying: "Custom rules the law" -"Phép vua thua lệ làng" [literally: the king's law yields to village customs]). Everyone in Vietnam wants to be buried in their village when they die.

Central and Eastern Europe

Slavic countries

Selo (Cyrillic: село; Polish: *wieś, sioło*) is a Slavic word meaning "village" in Bosnia and Herzegovina, Bulgaria, Croatia, Macedonia, Russia, Serbia, and Ukraine. For example there are numerous *sela* (plural of *selo*) called Novo Selo in Bulgaria, Croatia, Montenegro and others in Serbia, and Macedonia. In Slovenia, the word *selo* is used for very small villages (less than a thousand people) and in dialects; the Slovene word *vas* is used all over Slovenia.

Bulgaria

In Bulgaria, the different types of *Sela* vary from a small selo of 5 to 30 families to one of several thousand people. According to a 2002 census, in that year there were 2,385,000 Bulgarian citizens living in settlements classified as *villages*.[6] A 2004 Human Settlement Profile on Bulgaria[7] conducted by the United Nations Department of Economic and Social Affairs stated that:

Kovachevitsa, a village in southern Bulgaria

> The most intensive is the migration "city – city". Approximately 46% of all migrated people have changed their residence from one city to another. The share of the migration processes "village – city" is significantly less – 23% and "city – village" – 20%. The migration "village – village" in 2002 is 11%.[6]

It also stated that

> the state of the environment in the small towns and villages is good apart from the low level of infrastructure.[6]

In Bulgaria, it is becoming popular to visit villages for the atmosphere, culture, crafts, hospitality of the people and the surrounding nature. This is called *selski tourism* (Bulgarian: селски туризъм), meaning "village tourism".

Russia

In Russia, as of the 2010 Census, 26.3% of the country's population lives in rural localities;[8] down from 26.7% recorded in the 2002 Census.[8] Multiple types of rural localities exist, but the two most common are *derevnya* (деревня) and *selo* (село). Historically, the formal indication of status was religious: a city (*gorod*) had a cathedral, a *selo* had a church, while a *derevnya* had neither.

The village near Lake Baikal, Siberia

The lowest administrative unit of the Russian Empire, a *volost*, or its Soviet or modern Russian successor, a *selsoviet*, was typically headquartered in a *selo* and embraced a few neighboring villages.

Between 1926 and 1989, Russia's rural population shrank from 76 million people to 39 million, due to urbanization, collectivization, dekulakization, and the World War II losses, but has nearly stabilized since. During 1930–1937, mass starvation in Russia and other parts of the Soviet Union lead to the death of at least 14.5 million peasants (including 5–7 million in the Holodomor).[9]

Most Russian rural localities have populations of less than 200 people, and the smaller places take the brunt of depopulation: e.g., in 1959, about one half of Russia's rural population lived in villages of fewer than 500 people, while now less than one third does. In the 1960s–1970s, the depopulation of the smaller villages was driven by the central planners' drive to get the farm workers out of smaller, "prospect-less" hamlets and into the collective or state farms' main villages, with more amenities.[10]

Most Russian rural residents are involved in agricultural work, and it is very common for villagers to produce their own food. As prosperous urbanites purchase village houses for their second homes, Russian villages sometimes are transformed into dacha settlements, used mostly for seasonal residence.

The historically Cossack regions of Southern Russia and parts of Ukraine, with their fertile soil and absence of serfdom, had a rather different pattern of settlement from central and northern Russia. While peasants of central Russia lived in a village around the lord's manor, a Cossack family often lived on its own farm, called *khutor*. A number of such *khutors* plus a central village made up the administrative unit with a center in a *stanitsa* (Russian: станица; Ukrainian: станиця, *stanytsia*). Such *stanitsas* often with a few thousand residents, were usually larger than a typical *selo* in central Russia.

The term *aul/aal* is used to refer mostly Muslim-populated villages in Caucasus and Idel-Ural, without regard to the number of residents.

Ukraine

In Ukraine, a village, known locally as a *"selo"* (село), is considered the lowest administrative unit. Villages may have an individual administration (*silrada*) or a joint administration, combining two or more villages. Villages may also be under the jurisdiction of a city council (*miskrada*) or town council (*selyshchna rada*) administration.

There is, however, another smaller type of settlement which is designated in Ukrainian as a *selysche* (селище). This type of community is generally referred to in English as a "settlement". In comparison with an urban-type settlement, Ukrainian legislation does not have a concrete definition or a criterion to differentiate such settlements from villages. They represent a type of a small rural locality that might have once been a *khutir*, a fisherman's settlement, or a dacha. They are administered by a *silrada* (council) located in a nearby adjacent village. Sometimes the term *"selysche"* is also used in a more general way to refer to adjacent settlements near a bigger city, including urban-type settlements (*selysche miskoho typu*) and/or villages; however, ambiguity is often avoided in connection with urbanized settlements by referring to them using the three-letter abbreviation *smt* instead.

The largest Ukrainian village ("selo") Kosmach

The *khutir* (хутір) and *stanytsia* (станиця) are not part of the administrative division any longer, primarily due to collectivization. *Khutirs* were very small rural localities consisting of just few housing units and were sort of individual farms. They became really popular during the Stolypin reform in the early 20th century. During the collectivization, however, residents of such settlements were usually declared to be kulaks and had all their property confiscated and distributed to others (nationalized) without any compensation. The *stanitsa* likewise has not survived as an administrative term. The *stanitsa* was a type of a collective community that could include one or more settlements such as villages, *khutirs*, and others. Today,

stanitsa-type formations have only survived in Kuban (Russian Federation) where Ukrainians were resettled during the time of the Russian Empire.

Western & Southern Europe

United Kingdom

A village in the UK is a compact settlement of houses, smaller in size than a town, and generally based on agriculture or, in some areas, mining (such as Ouston, County Durham), quarrying or sea fishing.

The major factors in the type of settlement are location of water sources, organisation of agriculture and landholding, and likelihood of flooding. For example, in areas such as the Lincolnshire Wolds, the villages are often found along the spring line halfway down the hillsides, and originate as spring line settlements, with the original open field systems around the village. In northern Scotland, most villages are planned to a grid pattern located on or close to major roads, whereas in areas such as the Forest of Arden, woodland clearances produced small hamlets around village greens.[11] [12]

The main street of the village of Castle Combe, Wiltshire, England.

Some villages have disappeared (for example, deserted medieval villages), sometimes leaving behind a church or manor house and sometimes nothing but bumps in the fields.Some show archaeological evidence of settlement at three or four different layers, each distinct from the previous one. Clearances may have been to accommodate sheep or game estates, or enclosure, or may have resulted from depopulation, such as after the Black Death or following a move of the inhabitants to more prosperous districts. Other villages have grown and merged and often form hubs within the general mass of suburbia — such as Hampstead, London and Didsbury in Manchester. Many villages are now predominantly dormitory locations and have suffered the loss of shops, churches and other facilities.

For many British people, the village represents an ideal of Great Britain. Seen as being far from the bustle of modern life, it is represented as quiet and harmonious, if a little inward-looking. This concept of an unspoilt Arcadia is present in many popular representations of the village such as the radio serial *The Archers* or the best kept village competitions.[13]

Many villages in South Yorkshire, North Nottinghamshire, North East Derbyshire, County Durham, South Wales and Northumberland are known as pit villages. These (such as Murton, County Durham) grew from hamlets when the sinking of a colliery in the early 20th century resulted in a rapid growth in their population and the colliery owners built new housing, shops, pubs and churches. Some pit villages outgrew nearby towns by area and population; for example, Rossington in South Yorkshire came to have over four times more people than the nearby town of Bawtry. Some pit villages grew to become towns; for example, Maltby in South Yorkshire grew from 600 people in the 19th century[14] to over 17,000 in 2007.[15] Maltby was constructed under the auspices of the Sheepbridge Coal and Iron Company and included ample open spaces and provision for gardens.[16]

Bisley, Gloucestershire, a village in the Cotswolds

In the UK, the main historical distinction between a hamlet and a village was that the latter had a church,[1] and so usually was the centre of worship for an ecclesiastical parish. However, some civil parishes may contain more than one village. The typical village had a pub or inn, shops, and a blacksmith. But many of these facilities are now gone, and many villages are dormitories for commuters. The population of such settlements ranges from a few hundred people to around five thousand. A village is distinguished from a town in that:

- A village should not have a regular agricultural market, although today such markets are uncommon even in settlements which clearly are towns.
- A village does not have a town hall nor a mayor.
- If a village is the principal settlement of a civil parish, then any administrative body that administers it at parish level should be called a parish council or parish meeting, and not a town council or city council. However, some civil parishes have no functioning parish, town, or city council nor a functioning parish meeting. In Wales, where the equivalent of an English civil parish is called a Community, the body that administers it is called a Community Council. However, larger councils may elect to call themselves town councils.[17] Unlike Wales, Scottish community councils have no statutory powers.[18]
- There should be a clear green belt or open fields, as, for example, seen on aerial maps for Ouston surrounding its parish[19] borders. However this may not be applicable to urbanised villages: although these may not considered to be villages, they are often widely referred to as being so; an example of this is Horsforth in Leeds.

France

Same general definition as in the UK.

An independent association named *Les Plus Beaux Villages de France*, was created in 1982 to promote assets of small and picturesque French villages of quality heritage. As of 2008, 152 villages in France have been listed in "The Most Beautiful Villages of France".

Spain

Spain has plenty of little villages around its territory. The concept of village and country life is really present and usual in the North of the country (Atlantic area), especially in Galicia where villages are similar to English ones.

South of Barcelona is Spain's most romantic Mediterranean beach town, with a 2.5 km-long (1 1/2-mile) sandy beach and a promenade studded with flowers and palm trees. Sitges is a town with a rich connection to art; Picasso and Dalí both spent time here.[20] Mérida is an important Roman town with great tapas. Barcena Mayor (Cantabria) has houses that date back to the sixth century with simple two floors

Saint-Cirq-Lapopie (Lot) is one of "The Most Beautiful Villages in France".

constructions and rectangular form. Salamanca, an ancient Celtic town, is also a Renaissance city with striking architecture. Its sandstones buildings have a beautiful lustre giving the city the nickname, La Ciudad Dorado.[21]

Morella, Castellón is a medieval village located in the region of "Comunitat Valenciana" with huge castles with a rich renaissance history.[22] Rogueira pasturelands is one of the great ecological jewels of Galicia. Rivers, pools and springs abound in this verdant forest, as do underground water caves and caverns with a prehistoric past.[23] San Marti Vell is a charming small village well known for its Gothic spire. La Bisbal should be next on the list. The town is worth visiting for its Main Square and the castle. This Romanesque castle is situated in the middle of the town, giving it a romantic look. There is also Palafrugell, Palau-sator, Sant Julia and Sant Feliu de Boada. They are all very important because of their medieval patrimony. Castelló d'Empúries has 13th century Gothic churches. Angles also possesses outstanding medieval constructions throughout its village.[24] All villages have a church or hermitage.

Portugal

Villages are more usual in the northern and central regions and in the Alentejo. Most of them have a church and a "Casa do Povo" (people's house), where the village's summer **romarias** or religious festivities are usually held. Summer is also when many villages are host to a range of folk festivals and fairs, taking advantage of the fact that many of the locals who reside abroad tend to come back to their native village for the holidays.

Netherlands

In the flood prone districts of the Netherlands, villages were traditionally built on low man-made hills called terps before the introduction of regional dyke-systems. In modern days, the term *dorp* (lit. "village") is usually applied to settlements no larger than 20,000, though there's no official law regarding status of settlements in the Netherlands.

Middle East

Lebanon

Like France, villages in Lebanon are usually located in remote mountainous areas. The majority of villages in Lebanon retain their Aramaic names or are derivative of the Aramaic names, and this is because Aramaic was still in use in Mount Lebanon up to the 18th century.[25]

The main square of Saifi Village in Centre Ville, Beirut, Lebanon

Many of the Lebanese villages are a part of districts, these districts are known as "kadaa" which includes the districts of Baabda (Baabda), Aley (Aley), Matn (Jdeideh), Keserwan (Jounieh), Chouf (Beiteddine), Jbeil (Byblos), Tripoli (Tripoli), Zgharta (Zgharta / Ehden), Bsharri (Bsharri), Batroun (Batroun), Koura (Amioun), Miniyeh-Danniyeh (Minyeh / Sir Ed-Danniyeh), Zahle (Zahle), Rashaya (Rashaya), Western Beqaa (Jebjennine / Saghbine), Sidon (Sidon), Jezzine (Jezzine), Tyre (Tyre), Nabatiyeh (Nabatiyeh), Marjeyoun (Marjeyoun), Hasbaya (Hasbaya), Bint Jbeil (Bint Jbeil), Baalbek (Baalbek), and Hermel (Hermel).

The district of Danniyeh consists of thirty six small villages, which includes Almrah, Kfirchlan, Kfirhbab, Hakel al Azimah, Siir, Bakhoun, Miryata, Assoun, Sfiiri, Kharnoub, Katteen, Kfirhabou, Zghartegrein, Ein Qibil.

Danniyeh (known also as Addinniyeh, Al Dinniyeh, Al Danniyeh, Arabic: الضنيّة سير) is a region located in Miniyeh-Danniyeh District in the North Governorate of Lebanon. The region lies east of Tripoli, extends north as far as Akkar District, south to Bsharri District and Zgharta District and as far east as Baalbek and Hermel. Dinniyeh has an excellent ecological environment filled with woodlands, orchards and groves. Several villages are located in this mountainous area, the largest town being Sir Al Dinniyeh.

An example of a typical mountainous Lebanese village in Dannieh would be Hakel al Azimah which is a small village that belongs to the district of Danniyeh, situated between Bakhoun and Assoun's boundaries. It is in the centre of the valleys that lie between the Arbeen Mountains and the Khanzouh.

Syria

Syria contains a large number of villages that vary in size and importance, including the ancient, historical and religious villages, such as Ma'loula, Sednaya, and Brad (Mar Maroun's time). The diversity of the Syrian environments creates significant differences between the Syrian villages in terms of the economic activity and the method of adoption.

General view from Al-Annaze village, near Tartus, Syria

Villages in the south of Syria (Hauran, Jabal al-Druze), the north-east (the Syrian island) and the Orontes River basin depend mostly on agriculture, mainly grain, vegetables and fruits. Villages in the region of Damascus and Aleppo depend on trading. Some other villages, such as Marmarita depend heavily on tourist activity.

Mediterranean cities in Syria, such as Tartus and Latakia have similar types of villages. Mainly, villages were built in very good sites which had the fundamentals of the rural life, like water. An example of a Mediterranean Syrian village in Tartus would be al-Annaze, which is a small village that belongs to the area of al-Sauda. The area of al-Sauda is called a nahiya, which is a subdistrict.

Australasia and Oceania

Pacific Islands Communities on pacific islands were historically called villages by English speakers who traveled and settled in the area. Some communities such as several Villages of Guam continue to be called villages despite having large populations that can exceed 40,000 residents.

New Zealand The traditional Māori village was the pā, a fortified hill-top settlement. Tree-fern logs and flax were the main building materials. As in Australia (see below) the term is now used mainly in respect of shopping or other planned areas.

The village of Puamau on Hiva Oa, Marquesas Islands, French Polynesia

Australia The term village often is used in reference to small planned communities such as retirement communities or shopping districts, and tourist areas such as ski resorts. Small rural communities are usually known as townships. Larger settlements are known as towns.

South America

Argentina Usually set in remote mountainous areas, some also cater to winter sports and/or tourism, see: Uspallata, La Cumbrecita, Villa Traful and La Cumbre

North America

Canada

A Newfoundland fishing village

United States

Incorporated villages

In twenty[26] U.S. states, the term "village" refers to a specific form of incorporated municipal government, similar to a city but with less authority and geographic scope. However, this is a generality; in many states, there are villages that are an order of magnitude larger than the smallest cities in the state. The distinction is not necessarily based on population, but on the relative powers granted to the different types of municipalities and correspondingly, different obligations to provide specific services to residents.

A church in Newfane, Vermont

In some states such as New York, Wisconsin, or Michigan, a village is an incorporated municipality, usually, but not always, within a single town or civil township. Residents pay taxes to the village and town or township and may vote in elections for both as well. In some cases, the village may be coterminous with the town or township. There are also many villages which span the boundaries of more than one town or township, and some villages may even straddle county borders.

There is no limit to the population of a village in New York; Hempstead, the largest village in the state, has 55,000 residents, making it more populous than some of the state's cities. However, villages in the state may not exceed five square miles (13 km²) in area.

In the state of Wisconsin, a village is always legally separate from the towns that it has been incorporated from. The largest village is Menomonee Falls, which has over 32,000 residents.

Michigan and Illinois also have no set population limit for villages and there are many villages that are larger than cities in those states. The village of Arlington Heights, IL had 75,101 residents as of the 2010 census.

Villages in Ohio are often legally part of the township from which they were incorporated, although exceptions such as Hiram exist, in which the village is separate from the township.[27] They have no area limitations, but become cities if they grow a population of more than 5,000.[28]

In Maryland, a locality designated "Village of ..." may be either an incorporated town or a special tax district.[29] An example of the latter is the Village of Friendship Heights.

In states that have New England towns, a "village" is a center of population or trade, including the town center, in an otherwise sparsely-developed town or city — for instance, the village of Hyannis in the city of the Barnstable, Massachusetts.

Unincorporated villages

In many states, the term "village" is used to refer to a relatively small unincorporated community, similar to a hamlet in New York state. This informal usage may be found even in states that have villages as an incorporated municipality, although such usage might be considered incorrect and confusing.

Africa

Nigeria

Villages in Nigeria vary significantly because of cultural and geographical differences.

Northern Nigeria

In the North, villages were under traditional rulers long before the Jihad of Shaikh Uthman Bin Fodio and after the Holy War. At that time Traditional rulers used to have absolute power in their administrative regions. After Dan Fodio's Jihad in 1804,[30] political structure of the North became Islamic where emirs were the political, administrative and spiritual leaders of their people. These emirs appointed a number of people to assisted them in running the administration and that included villages. [31]

A village in Kaita, NigeriaKaita Nigeria

Every Hausa village was reigned by Magaji (Village head) who was answerable to his Hakimi (mayor) at town level. The Magaji also had his cabinet who assisted him rule his village efficiently, among whom was Mai-Unguwa (Ward Head). [32]

With the creation of Native Authority in Nigerian provinces, the autocratic power of village heads along with all other traditional rulers was subdued hence they ruled 'under the guidance of colonial officials'.[33]

Even though the constitution of the Federal Republic of Nigeria has not recognised the functions of traditional rulers, they still command respect in their villages[33] and political office holders liaise with them almost every time to reach people.

In Hausa language, village is called **ƙauye** and every local government area is made up of several small and large **ƙauyuka** (villages). For instance, Girka is a village in Kaita town in Katsina state in Nigeria. They have mud houses with thatched roofing though, like in most of villages in the North, zinc roofing is becoming a common sight.

Still in many villages in the North, people do not have access to portable water. So they fetch water from ponds and streams. Others are lucky to have wells within a walking distance. Women rush in the morning to fetch water in their clay pots from wells, boreholes and streams. However, government is now providing them with water bore holes.[34]

Electricity and GSM network are reaching more and more villages in the North almost everyday. So bad feeder roads may lead to remote villages with electricity and unstable GSM network.[35]

Southern Nigeria

Village dwellers in the Southeastern region lived separately in 'clusters of huts belonging to the patrilinage'.[36] As the rainforest region is dominated by Igbo speaking people, the villages are called **ime obodo** (inside town) in Igbo language. A typical large village might have a few thousand persons who shared the same market, meeting place and beliefs.

See also

- Global village
- Linear village
- Village green
- Village lock-up
- police village

Settlement types

- Dugout
- Fishing village
- Hamlet
- Microtown

Countries and localities

- Dhani and villages
- Dogon villages
- Hakka architecture
- Ksar
- List of villages in Europe by country
- Pueblo
- Sołectwo (rough equivalent in Poland)
- Ville

Developed environments

- Developed environments
- City
- Exurban
- Megalopolis
- Rural
- Suburban
- Urban area

Footnotes

[1] Dr Greg Stevenson, "What is a Village?" (http://www.bbc.co.uk/history/programmes/restoration/2006/exploring_brit_villages_01. shtml), *Exploring British Villages*, BBC, 2006, accessed 20 October 2009

[2] R.K. Bhatnagar. INDIA'S MEMBERSHIP OF ITER PROJECT (http://www.pibbng.kar.nic.in/feature1.pdf). PRESS INFORMATION BUREAU. GOVERNMENT OF INDIA, BANGALORE

[3] "Indian Census" (http://www.censusindia.gov.in/). Censusindia.gov.in. . Retrieved 2012-04-09.

[4] "Meriam-Webster Online" (http://www.m-w.com/dictionary/kampung). M-w.com. 2007-04-25. . Retrieved 2010-03-28.

[5] Geertz, Clifford. "Local Knowledge: Fact and Law in Comparative Perspective", pp. 167–234 in Geertz *Local Knowledge: Further Essays in Interpretive Anthropology*, NY: Basic Books. 1983.

[6] "Human Settlement Country Profile, Bulgaria (*2004*)" (http://www.un.org/esa/agenda21/natlinfo/countr/bulgaria/ Bulgariahumansettlement2003.PDF) (PDF). United Nations Department of Economic and Social Affairs. . Retrieved 2008-11-30.

[7] HUMAN SETTLEMENT COUNTRY PROFILE: BULGARIA (http://www.un.org/esa/agenda21/natlinfo/countr/bulgaria/ Bulgariahumansettlement2003.PDF). United Nations (2004)

[8] "Предварительные итоги Всероссийской переписи населения 2010 года[[Category:Articles containing non-English language text (http:// www.perepis-2010.ru/results_of_the_census/results-inform.php)] [Preliminary results of the 2010 All-Russian Population Census]"] (in Russian). *Всероссийская перепись населения 2010 года (2010 All-Russia Population Census)*. Federal State Statistics Service. 2011. . Retrieved February 9, 2012.

[9] Robert Conquest (1986) *The Harvest of Sorrow: Soviet Collectivization and the Terror-Famine.* Oxford University Press. ISBN 0-19-505180-7.

[10] "Российское село в демографическом измерении" (*Rural Russia measured demographically*) (http://demoscope.ru/weekly/2006/0253/tema04.php) **(Russian)**. This article reports the following census statistics:

Census year	1959	1970	1979	1989	2002
Total number of rural localities in Russia	294,059	216,845	177,047	152,922	155,289
Of them, with population 1 to 10 persons	41,493	25,895	23,855	30,170	47,089
Of them, with population 11 to 200 persons	186,437	132,515	105,112	80,663	68,807

[11] Wild, Martin Trevor (2004). *Village England: a social history of the countryside* (http://books.google.co.uk/books?id=M7AmyuGr5Y8C&printsec=frontcover). I.B.Tauris. p. 12. ISBN 978-1-86064-939-4. .

[12] Taylor, Christopher (1984). *Village and farmstead: A history of rural settlement in England* (http://books.google.co.uk/books?ei=NksHTpXqIo6t8QPE1ozADQ). G. Philip. p. 192. ISBN 978-0-540-01082-0. .

[13] OECD (2011). *OECD Rural Policy Reviews: England, United Kingdom 2011* (http://books.google.co.uk/books?id=bQCGWKfXJNMC&printsec=frontcover&source=gbs_ge_summary_r&cad=0). OECD Publishing. p. 237. .

[14] *The Parliamentary gazetteer of England and Wales* (http://books.google.co.uk/books?id=mxIQAAAAYAAJ&printsec=frontcover). **3**. A. Fullarton & Co.. 1851. p. 344. .

[15] "Maltby Ward" (http://www.rotherham.gov.uk/download/553/maltby_ward). Rotherham Metropolitan Borough Council. . Retrieved 2011-06-26.

[16] Baylies, Carolyn Louise (1993). *The history of the Yorkshire miners, 1881–1918* (http://books.google.co.uk/books?id=WEIOAAAAQAAJ&printsec=frontcover&source=gbs_ge_summary_r&cad=0). Routledge. .

[17] "National Statistics" (http://www.statistics.gov.uk/geography/parishes.asp). Statistics.gov.uk. . Retrieved 2010-03-28.

[18] "Portobello Community Council" (http://www.porty.org.uk/council/index.php). Porty.org.uk. . Retrieved 2010-03-28.

[19] "Ouston Parish Council" (http://parishes.durham.gov.uk/ouston/Pages/wherewelive.aspx). durham.gov.uk. .

[20] The Best Small Towns and Villages in Spain at Frommer's (http://www.frommers.com/destinations/spain/0242020855.html). Frommers.com. Retrieved on 2012-05-27.

[21] Spanish Towns and Villages (http://www.tourclare.com/spanishtownsandvillages.php). Tourclare.com. Retrieved on 2012-05-27.

[22] Top 10 Cities and Villages to Visit in Spain (http://www.travelthruspain.com/what-to-do/cities). Travelthruspain.com. Retrieved on 2012-05-27.

[23] Medieval villages in Galicia, Spain: O courel mountains. spain.info in English (http://www.spain.info/no/reportajes/sierra_de_o_courel_naturaleza_y_aldeas_medievales.html). Spain.info. Retrieved on 2012-05-27.

[24] Medieval Towns And Villages In Spain (http://travel.ezinemark.com/medieval-towns-and-villages-in-spain-16bc87fb4dc.html). Travel.ezinemark.com (2010-10-22). Retrieved on 2012-05-27.

[25] "A project proposal" (http://almashriq.hiof.no/lebanon/400/410/412/elies_project/glimse_of_yesterday.html). Almashriq.hiof.no. . Retrieved 2010-03-28.

[26] "Village" (http://www.websters-online-dictionary.org/definition/english/vi/village.html#Definitions). Websters-online-dictionary.org. . Retrieved 2010-03-28.

[27] "Detailed map of Ohio" (http://www2.census.gov/geo/maps/general_ref/cousub_outline/cen2k_pgsz/oh_cosub.pdf) (PDF). United States Census Bureau. 2000. . Retrieved 2010-03-28.

[28] "Ohio Revised Code Section 703.01(A)" (http://codes.ohio.gov/orc/703.01). . Retrieved 2010-03-28.

[29] 2002 Census of Governments, Individual State Descriptions (http://www.census.gov/prod/2005pubs/gc021x2.pdf) (PDF)

[30] Britannica Encyclopedia **The New Encyclopaedia Britannica** Vol. 6, 15th Edition. isbn 0-85229-961-3. Page 763

[31] Dr. Sani Abubakar Lugga. **The Great Province** published by Lugga Press Gidan Lugga, Kofar Marusa Road, Katsina Nigeria. ISBN 978-2105- 48 - 1. Page 43

[32] Dr. Sani Abubakar Lugga. **The Great Province** published by Lugga Press Gidan Lugga, Kofar Marusa Road, Katsina Nigeria. ISBN 978-2105- 48 - 1. Page 63

[33] A Johnson Ugoji Anyaele. **Comprehensive Government** A Johnson Publishers LTD. Surulere, Lagos. ISBN 978-2799-49-1. Page 123

[34] Katsina state government spends N4.2 billion on water supply project each year. http://www.peoplesdaily-online.com/index.php/news/news/special-report/7632-how-katsina-state-is-doing-so-much-with-so-little

[35] Nigerian Operator Expands Coverage. http://www.cellular-news.com/story/16837.php

[36] Villages in Igbo land prior to the colonial era. http://www.igboguide.org/HT-chapter10.htm

External links

- Types of villages (anthropogenic biomes) (http://www.ecotope.org/anthromes/v1/guide/villages/)

Obec

Obec (plural: obce) is the Czech and Slovak word for a municipality (both in the Czech Republic, Slovakia and abroad). The literal meaning of the word is "commune" or "community". It is the smallest administrative unit that is governed by elected representatives. Cities are also municipalities. The council is called "obecní zastupitelstvo/obecné zastupiteľstvo" or "zastupitelstvo města/mestské zastupiteľstvo" or "zastupitelstvo městyse", the office is called "obecní úřad/obecný úrad" or "úřad města/mestský úrad" or "úřad městyse". An obec can have its own flag and coat of arms. An obec is composed of one or more cadastral areas ("katastrální území/katastrálne územia"). Obec can have several settlements or parts whether villages or hamlets.[1]

Slovakia

There are 2891 obcí in Slovakia as of 2008.[2]

After meeting certain conditions such as population over 5000, being well accessible, having cultural or economical significance and having an urban style of settlement an obec can be declared a town ("mesto").[1]

Czech Republic

The number of municipalities (obcí) in the Czech Republic is 6250 (in 2010).

Whole area of the republic including mountains, forests and national parks is divided into municipalities (excepting military grounds). Smallest municipalities are shaped only by one small village, some by two or more villages or by the city and several villages. Mostly a municipality has the same name as the settlement where is the municipal office, but there are many exceptions: some municipalities have double name (Sedlec-Prčice, Brandýs nad Labem-Stará Boleslav), some municipalities have a name which have no its settlement (Orlické Podhůří) and some municipalities have the office in other part than in the "nominal".

The smallest Czech municipalities are Závist (0,42 km^2) and Karlova Studánka (0,46 km^2). The lowest number of the population has Vlkov (České Budějovice District) (16 inhabitants by ČSÚ, 2007) and Vysoká Lhota (22 inhabitants), many small municipalities have populations only of tens. Typical country municipality has hundreds of inhabitants.

The biggest are Prague (496,09 km^2, 1,3 mil. inh.), Brno (230,19 km^2, 405 thous.), Ostrava (214,22 km^2, 313 thous.) and by extent is the fourth Ralsko (170,23 km^2, including a former military area, only 2 thousands inhabitants).

Every city (město) or town (městys) is a "obec" (some separately, some along with attached villages). Not even biggest cities are diveded into more municipalites, but several biggest cities ("statute cities") can have self-governing subdivisions, so-called city parts (městská část) or city circuits (městský obvod) which have standing partly similar to small municipalities.

Some of municipalities have only basic status of municipality, some have extended competencies of delegated state administration (pověřená obec, obec s rozšířenou působností) for yourself and for surrounding lowly municipalities.

Město (a city) and městys (a town) is currently above all ceremonious honourary degree. Some municipalities have such titul for historic reasons, though they are small. Smallest cities are Přebuz (population of 69 inhabitants) and Loučná pod Klínovcem (87 inhabitants). The law makes it possible to restitute by request a status of city or town for every municipality which lost it (during communist period). But a municipality which want to acquire status of city (město) firstly, must have population over 3000 and the improvement in status is subject of assessment by chairman of the parliament. For městys (town) is no definite population. Biggest of cities are "statute cities" and can have self-governing subdivisions. A special type of municipality is the capital Prague, which has simultaneously a status of municipality and a status of region and which is treated by special law.

References

[1] "Zákona NR SR č. 369/1990 Zb. o obecnom zriadení" (http://www.zakonypreludi.sk/zz/2002-612) (in Slovak). www.zakonypreludi.sk. 2010-03-19. .

[2] "Urban and Municipal Statistics" (http://www.statistics.sk/mosmis/eng/run.html) (in Slovak). 2010-03-19. .

External links

- List of municipalities in Czech and Slovak Republic (http://municipality.euweb.cz/)

See also

- *okres*
- *opština*
- municipality
- List of cities in the Czech Republic
- List of municipalities in Slovakia

Human_settlement

Settlement, **locality** or **populated place** are general terms used in statistics, archaeology, geography, landscape history and other subjects for a permanent or temporary community in which people live or have lived, without being specific as to size, population or importance. A settlement can therefore range in size from a small number of dwellings grouped together to the largest of cities with surrounding urbanized areas. The term may include hamlets, villages, towns and cities.

The term is used internationally in the field of geospatial modeling, and in that context is defined as "a city, town, village, or other agglomeration of buildings where people live and work".[1]

The small town of Flora, Oregon in the United States is unincorporated, but is considered a populated place

A settlement conventionally includes its constructed facilities such as roads, enclosures, field systems, boundary banks and ditches, ponds, parks and woods, wind and water mills, manor houses, moats and churches.[2]

In geography

The United States Geological Survey (USGS) has a Geographic Names Information System that defines three classes of human settlement: "Populated Place", "Census" and "Civil".[3] The populated place is defined as a place or area with clustered or scattered buildings and a permanent human population (city, settlement, town, village or hamlet) referenced with geographic coordinates, which is "usually not incorporated and by definition has no legal boundaries".[3] The Populated Places may partially correspond to Civil records: "A political division formed for administrative purposes (borough, county, incorporated place, municipio, parish, town, township)."[3]

Taos Pueblo, an ancient pueblo belonging to a Taos speaking Native American tribe of Pueblo people. It is approximately 1000 years old and lies about 1 mile (1.6 km) north of the modern city of Taos, New Mexico.

In landscape history

Landscape history studies the form (morphology) of settlements – for example whether they are dispersed or nucleated. Urban morphology can thus be considered a special type of cultural-historical landscape studies. Settlements can be ordered by size, centrality or other factors to define a settlement hierarchy.

In statistics

Australia

Geoscience Australia defines a populated place as "a named settlement with a population of 200 or more persons."[4]

The Committee for Geographical Names in Australasia used the term localities for rural areas, while the Australian Bureau of Statistics uses the term "urban centres/localities" for urban areas.

Bulgaria

The Bulgarian Government publishes a National Register of Populated Places (NRPP).

Canada

The Canadian government uses the term "populated place" in the *Atlas of Canada*, but does not define it.[5] Statistics Canada uses the term localities for historical named locations.

Croatia

The Croatian Bureau of Statistics records population in units called settlements (*naselja*).

India

The Census Commission of India has a special definition of census towns.

Ireland

The Central Statistics Office of the Republic of Ireland has a special definition of census towns.

Russia

There are various types of inhabited localities in Russia.

Sweden

Statistics Sweden uses the term localities (*tätort*) for various densely populated places.

United Kingdom

The UK Department for Communities and Local Government uses the term "urban settlement" to denote an urban area when analysing census information.[6] The Registrar General for Scotland defines settlements as groups of one or more contiguous localities, which are determined according to population density and postcode areas. The Scottish settlements are used as one of several factors defining urban areas.[7]

United States

Populated places may be specifically defined in the context of censuses and be different from general-purpose administrative entities, such as "place" as defined by the U.S. Census Bureau or census-designated places.

Abandoned populated places

The term "Abandoned populated places" is a *Feature Designation Name* in databases sourced by the National Geospatial-Intelligence Agency[9] and GeoNames.[10]

Populated places can be abandoned. Sometimes the structures are still easily accessible, such as in a ghost town, and these may become tourist attractions. Some places that have the appearance of a ghost town, however, may still be defined as populated places by government entities.

Abandoned buildings in Kolmanskop, Namibia[8]

A town may become a ghost town because the economic activity that supported it has failed, because of a government action, such as the building of a dam that floods the town, or because of natural or human-caused disasters such as floods, uncontrolled lawlessness, or war. The term is sometimes used to refer to cities, towns, and neighborhoods that are still populated, but significantly less so than in years past.

See also

- Administrative division
- Lost city
- Requirements for permanent settlements
- List of Neolithic settlements
- Settlement geography

References

[1] Dutta, Biswanath; Fausto Giunchiglia and Vincenzo Maltese (2010). "A Facet-Based Methodology for Geo-Spatial Modeling" (http://
     eprints.biblio.unitn.it/archive/00001928/01/062.pdf). *GeoSpatial Semantics: 4th International Conference, GeoS 2011, Brest, France.*
     p. 143. .
[2] Medieval Settlement Research Group (http://www.britarch.ac.uk/msrg/msrgpolicy.htm)
[3] "Feature Class Definitions" (http://geonames.usgs.gov/pls/gnispublic/f?p=gnispq:8:3692675055014428). United States Geological
     Survey. . Retrieved July 26, 2011.
[4] "NTMS Specifications (250K & 100K): Populated Place" (http://www.ga.gov.au/mapspecs/250k100k/appendixA_files/Habitation.
     jsp#Habitation Populated Place Point). Australian Government. . Retrieved July 26, 2011.
[5] "Glossary Search Results" (http://atlas.nrcan.gc.ca/site/english/learningresources/glossary/results.html?term=). *Atlas of Canada.*
     Natural Resources Canada. . Retrieved July 26, 2011.
[6] Urban Settlement 2001 (http://www.communities.gov.uk/publications/planningandbuilding/urbansettlement2001)
[7] Scottish census information (http://www.gro-scotland.gov.uk/files1/the-census/key-stats-setloc-txt.pdf)
[8] "Maps of Kolmanskop - Namibia 2012" (http://en.mapatlas.org/Namibia/Abandoned_Populated_Place/Kolmanskop/3794/
     road_and_satellite_map). *Map Atlas - Google Maps based Atlas of the world* (http://en.mapatlas.org/). MapAtlas.org. 2012. . Retrieved
     February 11, 2012.
[9] "Feature Designation Code Lookup" (http://geonames.nga.mil/ggmagaz/feadesgsearchhtml.asp). *NGA: Geonames Search - OGC Viewer*
     (http://geonames.nga.mil/ggmaviewer/). Springfield, VA, USA: National Geospatial-Intelligence Agency. . Retrieved February 11, 2012.
[10] "GeoNames Feature Codes" (http://www.geonames.org/export/codes.html#P.PPLQ). *GeoNames* (http://www.geonames.org/).
     GeoNames. February 10, 2012. . Retrieved February 11, 2012.

Nymburk_District

<table>
<tr><td colspan="2" align="center">Nymburk District
Okres Nymburk</td></tr>
<tr><td colspan="2" align="center">— District —</td></tr>
<tr><td colspan="2" align="center">
Nymburk town</td></tr>
<tr><td colspan="2" align="center">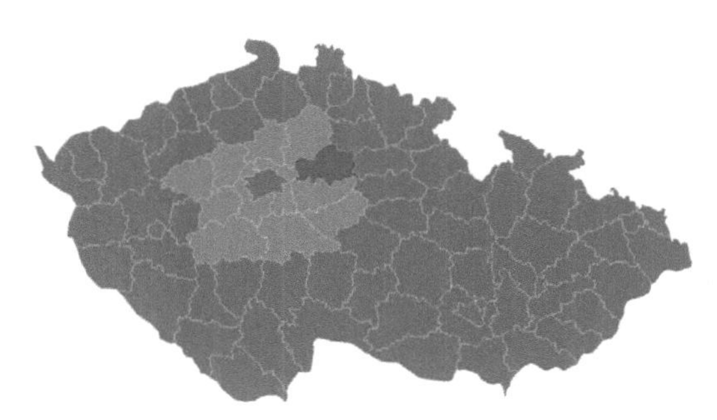
District location in the Central Bohemian Region within the Czech Republic</td></tr>
<tr><td>Country</td><td>Czech Republic</td></tr>
<tr><td>Region</td><td>Central Bohemian Region</td></tr>
<tr><td>Capital</td><td>Nymburk</td></tr>
<tr><td>Area</td><td></td></tr>
<tr><td>• Total</td><td>unknown operator: u'strong' sq mi (850.07 km^2)</td></tr>
<tr><td>Population (2007)</td><td></td></tr>
<tr><td>• Total</td><td>85674</td></tr>
<tr><td>• Density</td><td>unknown operator: u'strong'/sq mi (101/km^2)</td></tr>
</table>

Time zone	CET (UTC+1)
• Summer (DST)	CEST (UTC+2)

Nymburk District (*Okres Nymburk* in Czech) is a district (*okres*) within Central Bohemian Region (*Středočeský kraj*) of the Czech Republic. Its capital is city Nymburk.

Complete list of municipalities

Běrunice - Bobnice - Bříství - Budiměřice - Černíky - Čilec - Činěves - Dlouhopolsko - Dobšice - Dvory - Dymokury - Hořany - Hořátev - Hradčany - Hradištko - Hrubý Jeseník - Chleby - Choťánky - Chotěšice - Chrást - Chroustov - Jíkev - Jiřice - Jizbice - Kamenné Zboží - Kněžice - Kněžičky - Kolaje - Kostelní Lhota - Kostomlátky - Kostomlaty nad Labem - Košík - Kounice - Kouty - Kovanice - Krchleby - Křečkov - *Křinec* - Libice nad Cidlinou - *Loučeň* - **Lysá nad Labem** - Mcely - **Městec Králové** - Milčice - **Milovice** - Netřebice - Nový Dvůr - **Nymburk** - Odřepsy - Okřínek - Opočnice - Opolany - Oseček - Oskořínek - Ostrá - Pátek - Písková Lhota - Písty - **Poděbrady** - Podmoky - Přerov nad Labem - Rožďalovice - **Sadská** - Sány - Seletice - Semice - Senice - Sloveč - Sokoleč - Stará Lysá - Starý Vestec - Straky - Stratov - Třebestovice - Úmyslovice - Velenice - Velenka - Vestec - Vlkov pod Oškobrhem - Vrbice - Vrbová Lhota - Všechlapy - Vykáň - Záhornice - Zbožíčko - Zvěřínek - Žitovlice

Central_Bohemian_Region

<table>
<tr><td colspan="2" align="center">Central Bohemia
Středočeský kraj</td></tr>
<tr><td colspan="2" align="center">
Saint Barbara Cathedral in Kutná Hora</td></tr>
<tr><td align="center">
Flag</td><td align="center">
Coat of arms</td></tr>
<tr><td colspan="2" align="center">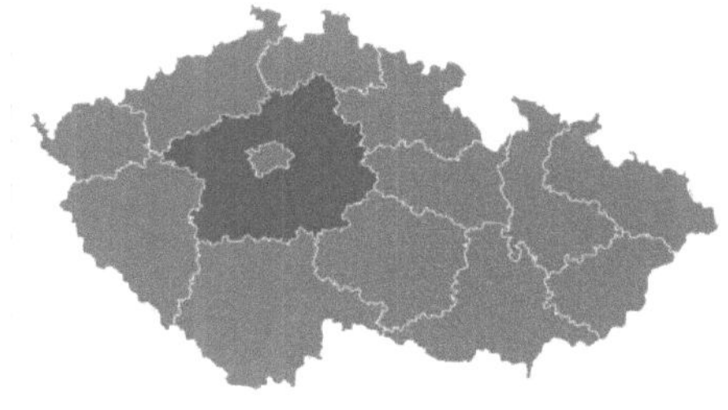</td></tr>
<tr><td>Country</td><td>Czech Republic</td></tr>
<tr><td>Government</td><td></td></tr>
<tr><td>• Governor</td><td>Josef Řihák</td></tr>
<tr><td>Area</td><td></td></tr>
<tr><td>• Total</td><td>11014 km^2 (unknown operator: u'strong' sq mi)</td></tr>
<tr><td>Highest elevation</td><td>865 m (2838 ft)</td></tr>
<tr><td>Population (03/2011)</td><td></td></tr>
<tr><td>• Total</td><td>1284629</td></tr>
<tr><td>• Density</td><td>unknown operator: u'strong'/km^2 (unknown operator: u'strong'/sq mi)</td></tr>
</table>

Time zone	CET (UTC+1)
• Summer (DST)	CEST (UTC+2)
ISO 3166-2	CZ-ST
Licence plate	S
NUTS code	CZ02
GDP per capita (PPS)	€ 17,200[1]
Website	http://www.kr-stredocesky.cz/ [2]

Central Bohemia (Czech: *Středočeský kraj*) is an administrative unit (Czech: *kraj*) of the Czech Republic, located in the central part of its historical region of Bohemia. Its administrative center is placed in the Czech capital Prague (Czech: *Praha*), which lies in the center of the region. The city is not, however, a part of it and creates a region of its own.

Districts

Benešov · Beroun · Kladno · Kolín · Kutná Hora · Mělník · Mladá Boleslav · Nymburk · Prague-East *(Praha-východ)* · Prague-West *(Praha-západ)* · Příbram · Rakovník

Cities and towns

- Beroun
- Benešov
- Brandýs nad Labem-Stará Boleslav
- Čáslav
- Dobřichovice
- Dobříš
- Kladno
- Kolín
- Kostelec nad Černými lesy
- Kralupy nad Vltavou
- Kutná Hora
- Mělník
- Mladá Boleslav
- Mnichovice (Prague-East District)
- Nymburk
- Poděbrady
- Příbram
- Rakovník
- Říčany
- Slaný
- Vlašim

Castles

- Karlštejn Castle
- Kokořín Castle
- Konopiště
- Křivoklát Castle
- Lány

References

[1] "GDP per inhabitant in 2006 ranged from 25% of the EU27 average in Nord-Est in Romania to 336% in Inner London" (http://epp.eurostat.
   ec.europa.eu/pls/portal/docs/PAGE/PGP_PRD_CAT_PREREL/PGE_CAT_PREREL_YEAR_2009/
   PGE_CAT_PREREL_YEAR_2009_MONTH_02/1-19022009-EN-AP.PDF). Eurostat. .
[2] http://www.kr-stredocesky.cz/

External links

- Official website (http://www.kr-stredocesky.cz/home.asp?lang=en)
- Region statistics (http://www.czech.cz/index.php?section=1&menu=5&action=text&id=149)

Bříství

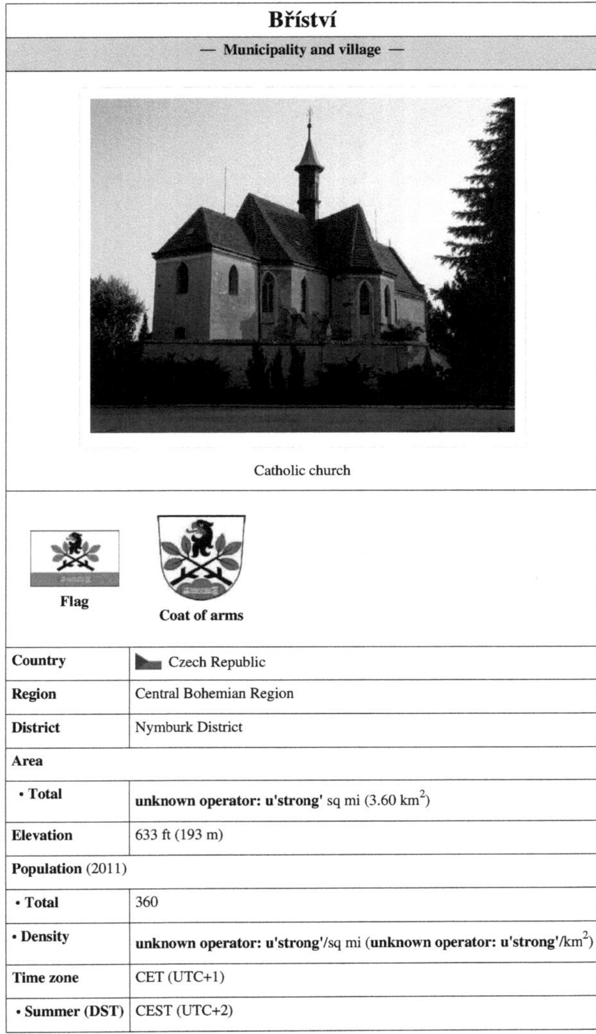

<table>
<tr><td colspan="2" align="center">Bříství</td></tr>
<tr><td colspan="2" align="center">— Municipality and village —</td></tr>
<tr><td colspan="2" align="center">Catholic church</td></tr>
<tr><td align="center">Flag</td><td align="center">Coat of arms</td></tr>
<tr><td>Country</td><td>Czech Republic</td></tr>
<tr><td>Region</td><td>Central Bohemian Region</td></tr>
<tr><td>District</td><td>Nymburk District</td></tr>
<tr><td>Area</td><td></td></tr>
<tr><td>• Total</td><td>unknown operator: u'strong' sq mi (3.60 km²)</td></tr>
<tr><td>Elevation</td><td>633 ft (193 m)</td></tr>
<tr><td>Population (2011)</td><td></td></tr>
<tr><td>• Total</td><td>360</td></tr>
<tr><td>• Density</td><td>unknown operator: u'strong'/sq mi (unknown operator: u'strong'/km²)</td></tr>
<tr><td>Time zone</td><td>CET (UTC+1)</td></tr>
<tr><td>• Summer (DST)</td><td>CEST (UTC+2)</td></tr>
</table>

Bříství is a village and municipality in Nymburk District in the Central Bohemian Region of the Czech Republic.

The European route E67 near Bříství

Notes

- *This article was initially translated from the Czech Wikipedia.*

Bohemia

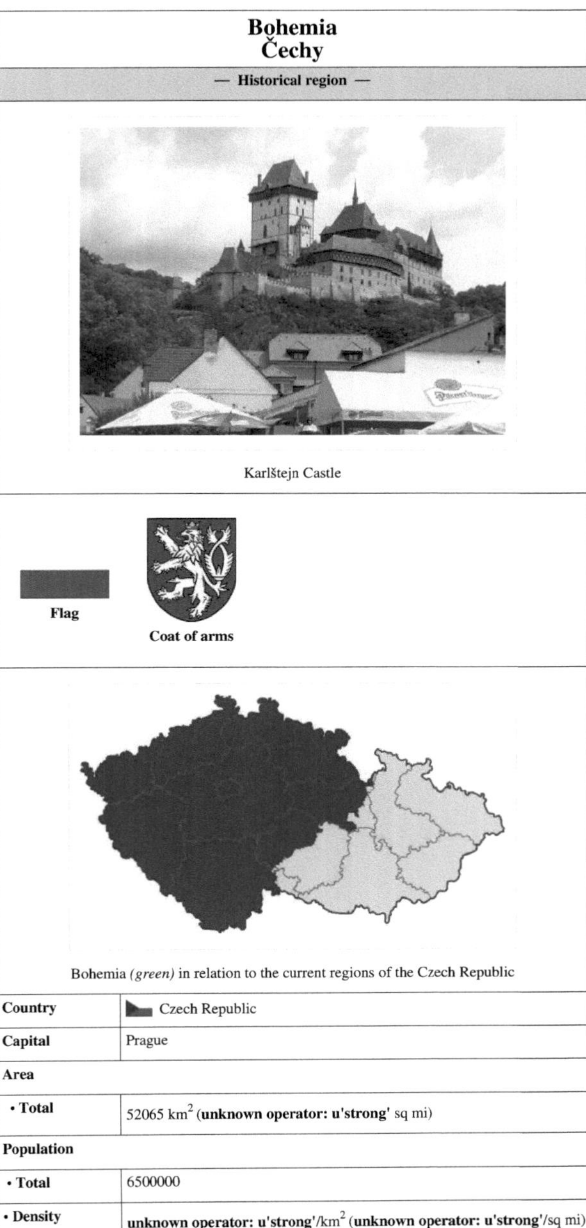

Country	Czech Republic
Capital	Prague
Area	
• **Total**	52065 km^2 (**unknown operator: u'strong'** sq mi)
Population	
• **Total**	6500000
• **Density**	**unknown operator: u'strong'**/km^2 (**unknown operator: u'strong'**/sq mi)

Time zone	CET (UTC+1)
• **Summer (DST)**	CEST (UTC+2)

Bohemia (Czech: *Čechy*;[1] German: *Böhmen*; Polish: *Czechy*; French: *Bohême*; Latin: *Bohemia*) is a historical region in central Europe, occupying the western two-thirds of the traditional Czech Lands. It is located in the contemporary Czech Republic with its capital in Prague. In a broader meaning, it often refers to the entire Czech territory, including Moravia and Czech Silesia,[2] especially in historical contexts, such as the Kingdom of Bohemia. Bohemia is a historic country of central Europe that was a kingdom in the Holy Roman Empire and subsequently a province in the Habsburgs' Austrian Empire. Bohemia was bounded on the south by Upper and Lower Austria, on the west by Bavaria, on the north by Saxony and Lusatia, on the northeast by Silesia, and on the east by Moravia. From 1918 to 1939 and from 1945 to 1992 it was part of Czechoslovakia, and since 1993 it has formed much of the Czech Republic.[3]

Bohemia has an area of 52,065 km² and today is home to approximately 6 million of the Czech Republic's 10.3 million inhabitants. It is bordered by Germany to the west, Poland to the northeast, the historical region of Moravia to the east, and Austria to the south. Bohemia's borders are marked with mountain ranges such as the Bohemian Forest, the Ore Mountains, and the Krkonoše (Giant Mountains), the highest within the Sudeten mountain range.

Etymology

In the 2nd century BC, the Romans were competing for dominance in northern Italy, with various peoples including the Boii. The Romans defeated the Boii at the Battle of Placentia (194 BC) and the Battle of Mutina (193 BC). After this, many of the Boii retreated north across the Alps.[4]

Roman authors refer to the area they invaded as **Boihaemum**, the earliest mention[4] being in Tacitus' *Germania* 28[5] (written at the end of the 1st century AD). The name appears to include the tribal name *Boi-* plus the Germanic element **xaim-* "home" (whence Gothic *haims*, German *Heim*, English *home*). This Boihaemum included parts of southern Bohemia as well as parts of Bavaria (whose name also seems to derive from the tribal name Boii) and Austria. The Czech name "Čechy" is derived from the name of the Slavic tribe of Czechs which settled down in this area in 6th or 7th century.

History

Further information: History of the Czech lands and History of Czechoslovakia

Ancient Bohemia

In the late 2nd century BC, many of the Boii tribe, after defeat at Roman hands, fled north across the Alps from northern Italy into an area called **Boihaemum** by the Romans, which included the southern part of present-day Bohemia.

The western half was conquered and settled from the 1st century BC by Germanic (probably Suebic) peoples including the Marcomanni; the elite of some Boii then migrated west to modern Switzerland and southeastern Gaul. Those Boii that remained in the eastern part were eventually absorbed by the Marcomanni. Part of the Marcomanni, renamed the Bavarians (Baiuvarii), later migrated to the southwest. Although the leading tribes changed, there was a large degree of continuity in the actual population, and at no time was there a wholesale depopulation or change in ethnic stock.

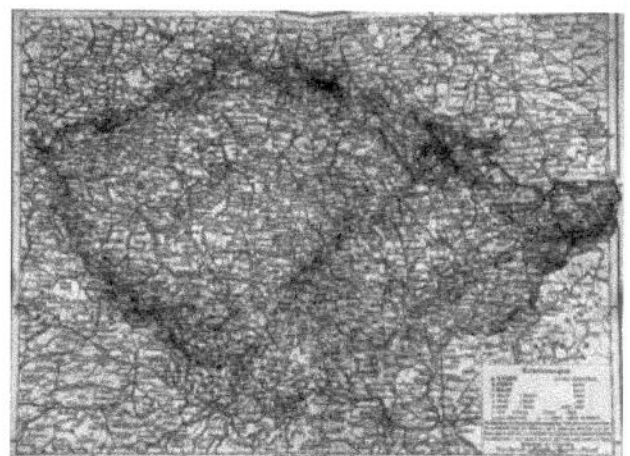

Historical map with Bohemia proper outlined in pink, Moravia in yellow, and Austrian Silesia in orange.

After the Bavarian emigration, Bohemia was partially repopulated around the 6th century, as part of the territory often crossed during the Migration Period by Germanic and major Slavic tribes, precursors of today's Czechs, though the exact amount of Slavic immigration is a subject of debate. The Slavic influx was divided into two or three waves. The first wave came from the southeast and east, when the Germanic Lombards left Bohemia (c. 568 AD). Soon after, from the 630s to 660s, the territory was taken by Samo's tribal confederation. His death marked the end of the old "Slavonic" confederation, the second attempt to establish such a Slavonic union after Carantania in Carinthia.

Other sources (*Descriptio civitatum et regionum ad septentrionalem plagam Danubii*, Bavaria, 800–850) divide the population of Bohemia at this time into the Merehani, Marharaii, Beheimare (Bohemani) and Fraganeo. (The suffix *-ani* or *-ni* means "people of-"). The great tribes of *Dudleb*, *Lemuz* and *Charvat* are missing from this list, which shows a linguistic and cultural shift from Sarmatian in favor of Slavonic dialects, a common occurrence in nomadic immigrations. Christianity first appeared in the early 9th century, but only became dominant much later, in the 10th or 11th century.

The 9th century was crucial for the future of Bohemia. The manorial system sharply declined, as it did in Bavaria. The influence of the central *Fraganeo-Czechs* grew, as a result of the important cultic centre in their territory. They were Slavic-speaking and thus contributed to the transformation of diverse neighbouring populations into a new nation named and led by them with a united *slavic* ethnic consciousness.[6]

Přemysl dynasty

The Coat of arms of the Bohemian King and Kingdom.

Initially, Bohemia was a part of Greater Moravia. The latter, which had been weakened by years of internal conflict and constant warfare, ultimately succumbed and fragmented due to the continual incursions of the invading nomadic Magyars. However, Bohemia's initial incorporation into the Moravian Empire resulted in the extensive Christianization of the population. A native monarchy arose to the throne, and Bohemia came under the rule of the Přemyslid dynasty, which would rule the Czech lands for the next several hundred years.

The Přemyslids secured their frontiers from the remnant Asian interlocurs, after the collapse of the Moravian state, by entering into a state of semi-vassalage to the Frankish rulers. This alliance was facilitated by Bohemia's conversion to Christianity, in the 9th century. Continuing close relations were developed with the East Frankish kingdom, which devolved from the Carolingian Empire, into East Francia, eventually becoming the Holy Roman Empire.

After a decisive victory of the Holy Roman Empire and Bohemia over invading Magyars in the 955 Battle of Lechfeld, Boleslaus I of Bohemia was granted the March of Moravia by German emperor Otto the Great. Bohemia would remain a largely autonomous state under the Holy Roman Empire for several decades. The jurisdiction of the Holy Roman Empire was definitively reasserted when Jaromír of Bohemia was granted fief of the Kingdom of Bohemia by Emperor King Henry II of the Holy Roman Empire, with the promise that he hold it as a vassal once he re-occupied Prague with a German army in 1004, ending the rule of Boleslaw I of Poland.

The first to use the title of "King of Bohemia" were the Přemyslid dukes Vratislav II (1085) and Vladislav II (1158), but their heirs would return to the title of duke. The title of king became hereditary under Ottokar I (1198). His grandson Ottokar II (king from 1253–1278) conquered a short-lived empire which contained modern Austria and Slovenia. The mid-13th century saw the beginning of substantial German immigration as the court sought to replace losses from the brief Mongol invasion of Europe in 1241. Germans settled primarily along the northern, western, and southern borders of Bohemia, although many lived in towns throughout the kingdom.

Luxembourg dynasty

The House of Luxembourg accepted the invitation to the Bohemian throne with the marriage to the Premsylid heiress, Elizabeth and the crowning subsequent of John I of Bohemia in 1310. His son, Charles IV became King of Bohemia in 1346. He founded Charles University in Prague, central Europe's first university, two years later. His reign brought Bohemia to its peak both politically and in total area, resulting in his being the first King of Bohemia to also be elected as Holy Roman Emperor. Under his rule the Bohemian crown controlled such diverse lands as Moravia, Silesia, Upper Lusatia and Lower Lusatia, Brandenburg, an area around Nuremberg called New Bohemia, Luxembourg, and several small towns scattered around Germany.

Hussite Bohemia

During the ecumenical Council of Constance in 1415, Jan Hus, the rector of Charles University and a prominent reformer and religious thinker, was sentenced to be burnt at the stake as a heretic. The verdict was passed despite the fact that Hus was granted formal protection by Emperor Sigismund of Luxembourg prior to the journey. Hus was invited to attend the council to defend himself and the Czech positions in the religious court, but with the emperor's approval, he was executed on 6 July 1415. The execution of Hus, as well as five consecutive papal crusades against followers of Hus, forced the Bohemians to defend themselves. Their defense and rebellion against Roman Catholics became known as the Hussite Wars.

The uprising against imperial forces was led by a former mercenary, Jan Žižka of Trocnov. As the leader of the Hussite armies, he used innovative tactics and weapons, such as howitzers, pistols, and fortified wagons, which were revolutionary for the time, and established Žižka as a great general who never lost a battle.

After Žižka's death, Prokop the Great took over the command for the army, and under his lead the Hussites were victorious for another ten years, to the sheer terror of Europe. The Hussite cause gradually splintered into two main factions, the moderate Utraquists and the more fanatic Taborites. The Utraquists began to lay the groundwork for an agreement with the Catholic Church and found the more radical views of the Taborites distasteful. Additionally, with general war weariness and yearning for order, the Utraquists were able to eventually defeat the Taborites in the Battle of Lipany in 1434. Sigismund said after the battle that "only the Bohemians could defeat the Bohemians."

Despite an apparent victory for the Catholics, the Bohemian Utraquists were still strong enough to negotiate freedom of religion in 1436. This happened in the so-called Basel Compacts, declaring peace and freedom between Catholics and Utraquists. It would only last for a short period of time, as Pope Pius II declared the Basel Compacts to be invalid in 1462.

In 1458, George of Podebrady was elected to ascend to the Bohemian throne. He is remembered for his attempt to set up a pan-European "Christian League", which would form all the states of Europe into a community based on religion. In the process of negotiating, he appointed Leo of Rozmital to tour the European courts and to conduct the talks. However, the negotiations were not completed, because George's position was substantially damaged over time by his deteriorating relationship with the Pope.

Habsburg Monarchy

After the death of King Louis II of Hungary and Bohemia in the Battle of Mohács in 1526, Archduke Ferdinand of Austria became King of Bohemia and the country became a constituent state of the Habsburg Monarchy.

Bohemia enjoyed religious freedom between 1436 and 1620, and became one of the most liberal countries of the Christian world during that period. In 1609, Holy Roman Emperor Rudolph II who made Prague again the capital of the Empire at the time, himself a Roman Catholic, was moved by the Bohemian nobility to publish *Maiestas Rudolphina*, which confirmed the older *Confessio Bohemica* of 1575.

After Emperor Ferdinand II began oppressing the rights of Protestants in Bohemia, the resulting Bohemian Revolt led to outbreak of the Thirty Years' War in 1618. Elector Frederick V of the Electorate of the Palatinate, a Protestant, was elected by the Bohemian nobility to replace Ferdinand on the Bohemian throne, and was known as the Winter King. Frederick's wife, the popular Elizabeth Stuart and subsequently Elizabeth of Bohemia, known as the Winter Queen or Queen of Hearts, was the daughter of King James I of England. However, after Frederick's defeat in the Battle of White Mountain in 1620, 27 Bohemian estates leaders together with Jan Jesenius, rector of the Charles University of Prague were executed on the Prague's Old Town Square on 21 June 1621 and the rest were exiled from the country; their lands were then given to Catholic loyalists (mostly of Bavarian and Saxon origin), this ended the pro-reformation movement in Bohemia and also ended the role of Prague as ruling city of the Holy Roman Empire.

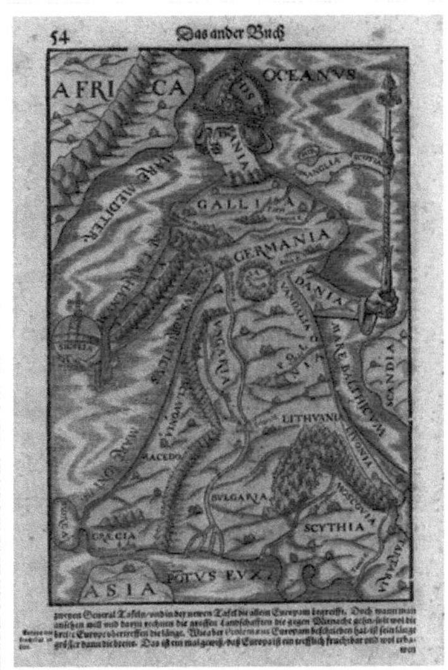

Bohemia as the heart of *Europa regina*, 1570

Until the so-called "renewed constitution" of 1627, the German language was established as a second official language in the Czech lands. The Czech language remained the first language in the kingdom. Both German and Latin were widely spoken among the ruling classes, although German became increasingly dominant, while Czech was spoken in much of the countryside.

The formal independence of Bohemia was further jeopardized when the Bohemian Diet approved administrative reform in 1749. It included the indivisibility of the Habsburg Empire and the centralization of rule; this essentially meant the merging of the Royal Bohemian Chancellery with the Austrian Chancellery.

At the end of the 18th century, the Czech National Revival movement, in cooperation with part of the Bohemian aristocracy, started a campaign for restoration of the kingdom's historic rights, whereby the Czech language was to replace German as the language of administration. The enlightened absolutism of Joseph II and Leopold II, who introduced minor language concessions, showed promise for the Czech movement, but many of these reforms were later rescinded. During the Revolution of 1848, many Czech nationalists called for autonomy for Bohemia from Habsburg Austria, but the revolutionaries were defeated. The old Bohemian Diet, one of the last remnants of the independence, was dissolved, although the Czech language experienced a rebirth as romantic nationalism developed among the Czechs.

In 1861, a new elected Bohemian Diet was established. The renewal of the old Bohemian Crown (Kingdom of Bohemia, Margraviate of Moravia, and Duchy of Silesia) became the official political program of both Czech liberal politicians and the majority of Bohemian aristocracy ("state rights program"), while parties representing the German minority and small part of the aristocracy proclaimed their loyalty to the centralistic Constitution (so-called

"Verfassungstreue"). After the defeat of Austria in the Austro-Prussian War in 1866, Hungarian politicians achieved the Austro-Hungarian Compromise of 1867, ostensibly creating equality between the Austrian and Hungarian halves of the empire. An attempt by the Czechs to create a tripartite monarchy (Austria-Hungary-Bohemia) failed in 1871. However, the "state rights program" remained the official platform of all Czech political parties (except for social democrats) until 1918.

Twentieth century

After World War I, Bohemia (as the biggest and most populated land) became the core of the newly-formed country of Czechoslovakia, which combined Bohemia, Moravia, Austrian Silesia, Upper Hungary (present-day Slovakia) and Carpathian Ruthenia into one state. Under its first president, Tomáš Masaryk, Czechoslovakia became a liberal democratic republic but serious issues emerged regarding the Czech majority's relationship with the native German and Hungarian minorities.

Bohemia (westernmost area) within Czechoslovakia between 1928–38

The Bohemian town of Karlovy Vary

Following the Munich Agreement in 1938, the border regions of Bohemia historically inhabited predominantly by ethnic Germans (the Sudetenland) were annexed to Nazi Germany; this was the only time in Bohemian history that its territory was politically divided. The remnants of Bohemia and Moravia were then annexed by Germany in 1939, while the Slovak lands became the separate Slovak Republic, a puppet state of Nazi Germany. From 1939 to 1945 Bohemia, (without the Sudetenland), together with Moravia formed the German Protectorate of Bohemia and Moravia (*Reichsprotektorat Böhmen und Mähren*). Any open opposition to German occupation was brutally suppressed by the Nazi authorities and many Czech patriots were executed as a result. After World War II ended in 1945, the vast majority of remaining Germans were expelled by force by the order of the re-established Czechoslovak central government, based on the Potsdam Agreement, and their property was confiscated by the Czech authorities. This severely depopulated the area and from this moment on locales were only referred to in their Czech equivalents regardless of their previous demographic makeup. In 1946, per the Potsdam Agreement, and under the stipulation that it be placed "under Polish administration" the post war Communist Party backed by the Soviet Union re-established Czechoslovakia. The Party won the most votes in free elections but not a simple majority. Klement Gottwald, the communist leader, became Prime Minister of a coalition government. In February 1948 the non-communist members of the government resigned in protest against arbitrary measures by the communists and their Soviet protectors in many of the state's institutions. Gottwald and the communists responded with a coup d'état and installed a pro-Soviet authoritarian state.

Beginning in 1949, Bohemia ceased to be an administrative unit of Czechoslovakia, as the country was divided into administrative regions. Between 1949 and 1989 Czechoslovakia (from 1960 officially called Czechoslovak Socialistic Republic) became a Soviet satellite even though there was not a Soviet army present until 1968 (interestingly enough, surrounding countries including Eastern Austria were occupied by the Red Army) when

Czechoslovak Communist Party started to reform and democratize itself. This "Prague Spring" process was stopped abruptly by an invasion of 'brotherly' armies of Warsaw Pact in August 1968. "Temporary stationing" of Soviet army following the invasion ended in 1991. In 1989, Agnes of Bohemia became the first saint from a Central European country to be canonized by Pope John Paul II before the "Velvet Revolution" later that year. After the dissolution of Czechoslovakia in 1993 (the "Velvet Divorce"), the territory of Bohemia became part of the new Czech Republic.

The Czech constitution from 1992 refers to the "citizens of the Czech Republic in Bohemia, Moravia and Silesia" and proclaims continuity with the statehood of the Bohemian Crown. Bohemia is not currently an administrative unit of the Czech Republic. Instead, it is divided into the Prague, Central Bohemia, Plzeň, Karlovy Vary, Ústí nad Labem, Liberec, and Hradec Králové Regions, as well as parts of the Pardubice, Vysočina, South Bohemian and South Moravian Regions.

See also

- Bohemianism
- Crown of Bohemia
- German Bohemia
- History of the Czech lands
- Lech, Czech and Rus
- List of rulers of Bohemia
- Sudetenland
- Bohemia at the 1908 Summer Olympics

References

[1] There is no distinction in the Czech language between adjectives referring to Bohemia and to the Czech Republic; i.e. *český* means both *Bohemian* and *Czech*.

[2] The Columbia Encyclopedia, Sixth Edition. 2001–05

[3] "Bohemia" (http://www.britannica.com/EBchecked/topic/71528/Bohemia). . Retrieved June 2, 2012.

[4] Collis, John. *The Celts: Origins, Myth and Inventions*. Tempus Publishing, 2003. ISBN 0-7524-2913-2

[5] http://www.thelatinlibrary.com/tacitus/tac.ger.shtml#28

[6] Petr Charvát: "Zrod Českého státu" [Origin of the Bohemian State], March 2007, ISBN 80-7021-845-2, in Czech

External links

- Bohemia (http://www.czech.cz/)

Czech_Republic

<table>
<tr><td colspan="2" align="center">Czech Republic
Česká republika</td></tr>
<tr><td colspan="2"> </td></tr>
<tr><td colspan="2">Motto: "Pravda vítězí" (Czech)
"Truth prevails"</td></tr>
<tr><td colspan="2">Anthem: Kde domov můj? (Czech) [a]
"Where is my homeland?"</td></tr>
<tr><td colspan="2">

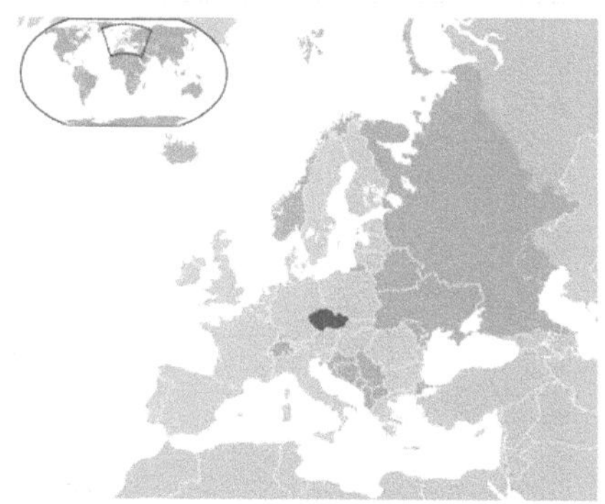

Location of Czech Republic(dark green)

— in Europe(green & dark grey)
— in the European Union(green) — [Legend]

</td></tr>
<tr><td>Capital
and largest city</td><td>Prague (Praha)
50°05′N 14°28′E</td></tr>
<tr><td align="center">Official languages</td><td>Czech[1]</td></tr>
<tr><td align="center">Officially recognized minority languages[2] [3]</td><td>Slovak, German, Polish, Bulgarian, Croatian, Greek, Hungarian, Romani, Russian, Rusyn, Serbian, Ukrainian</td></tr>
<tr><td>Ethnic groups (2011[4])</td><td>

- 63.7% Czechs
- 4.9% Moravians
- 1.4% Slovaks
- 29.9% others and unspecified

</td></tr>
<tr><td align="center">Demonym</td><td>Czech</td></tr>
<tr><td align="center">Government</td><td>Parliamentary republic</td></tr>
<tr><td>- President</td><td>Václav Klaus</td></tr>
<tr><td>- Prime Minister</td><td>Petr Nečas</td></tr>
<tr><td align="center">Legislature</td><td>Parliament</td></tr>
</table>

-	Upper house	Senate
-	Lower house	Chamber of Deputies
Formation		
-	Principality of Bohemia	c. 870
-	Kingdom of Bohemia	1198
-	Czechoslovakia	28 October 1918
-	Czech Socialist Republic	1 January 1969
-	Czech Republic	1 January 1993
Area		
-	Total	78,866 km^2 (116th) 30,450 sq mi
-	Water (%)	2
Population		
-	Sep 2012 estimate	10,513,209[5] (81st)
-	2011 census	10,436,560[6]
-	Density	134/km^2 (84th) 341/sq mi
GDP (PPP)		2012 estimate
-	Total	$286.676 billion[7]
-	Per capita	$27,165[7]
GDP (nominal)		2012 estimate
-	Total	$193.513 billion[7]
-	Per capita	$18,337[7]
Gini (2008)		26 (low / 4th)
HDI (2010)		▲ 0.865[8] (very high / 27th)
	Currency	Czech koruna (CZK)
	Time zone	CET (UTC+1)
-	Summer (DST)	CEST (UTC+2)
	Drives on the	right
	Calling code	+420^c
	ISO 3166 code	CZ
	Internet TLD	.cz^b
a.	The question is rhetorical, implying "those places where my homeland lies".	
b.	Also .eu, shared with other European Union member states.	
c.	Code 42 was shared with Slovakia until 1997.	

The **Czech Republic** (🔊 ⁱ/ˈtʃɛk/ *CHEK*;[9] Czech: *Česká republika*, pronounced [ˈtʃɛskaː ˈrɛpuˌblɪka] (🔊 listen), short form *Česko* Czech pronunciation: [ˈtʃɛsko]), is a landlocked country in Central Europe. The country is bordered by Germany to the west, Austria to the south, Slovakia to the east and Poland to the north. Its capital and largest city, with 1.3 million inhabitants, is Prague. The Czech Republic includes the historical territories of Bohemia and Moravia and a small part of Silesia.

The Czech state, formerly known as **Bohemia**, was formed in the late 9th century as a small duchy around Prague, at that time under the dominance of the powerful Great Moravian Empire. After the fall of the Empire in 907, the centre of power was transferred from Moravia to Bohemia, under the Přemyslids. Since 1002 it was formally recognized as part of Holy Roman Empire.[10] [11] In 1212 the duchy was raised to a kingdom and during the rule of Přemyslid dukes/kings and their successors, the Luxembourgs, the country reached its greatest territorial extent (13th–14th century). During the Hussite wars the kingdom faced economic embargoes and crusades from all over Europe. Following the Battle of Mohács in 1526, the Kingdom of Bohemia was gradually integrated into the Habsburg monarchy as one of its three principal parts, alongside the Archduchy of Austria and the Kingdom of Hungary. The Bohemian Revolt (1618–20) lost in the Battle of White Mountain, led to the further centralization of the monarchy including forced recatholization and Germanization. With the dissolution of the Holy Roman Empire in 1806, the Bohemian kingdom became part of the Austrian Empire. In the 19th century the Czech lands became the industrial powerhouse of the monarchy and the core of the Republic of Czechoslovakia which was formed in 1918, following the collapse of the Austro-Hungarian Empire after World War I. After 1933, Czechoslovakia remained the only democracy in central and eastern Europe.

After the Munich Agreement, Polish annexation of Zaolzie and German occupation of Czechoslovakia and the consequent disillusion with the Western response and gratitude for the liberation of the major portion of Czechoslovakia by the Red Army, the Communist Party of Czechoslovakia won the majority in the 1946 elections. In the 1948 coup d'état, Czechoslovakia became a communist-ruled state. In 1968, the increasing dissatisfaction culminated in attempts to reform the communist regime. The events, known as the Prague Spring of 1968, ended with an invasion by the armies of the Warsaw Pact countries (with the exception of Romania); the troops remained in the country until the 1989 Velvet Revolution, when the communist regime collapsed. On 1 January 1993, Czechoslovakia peacefully dissolved into its constituent states, the Czech Republic and the Slovak Republic.

The Czech Republic is the first former member of the Comecon to achieve the status of a developed country according to the World Bank.[12] In addition, the country has the highest human development in Central and Eastern Europe,[13] ranking as a "Very High Human Development" nation. It is also ranked as the third most peaceful country in Europe and most democratic and healthy (by infant mortality) country in the region. It is a pluralist multi-party parliamentary representative democracy, a member of the European Union, NATO, the OECD, the OSCE, the Council of Europe and the Visegrád Group.

Etymology

The traditional English name "Bohemia" derives from Latin "Boiohaemum", which means "home of the Boii". The current name comes from the endonym *Čech*, borrowed through Polish and spelt accordingly.[14] [15] The name comes from the Slavic tribe (Czechs, Czech: *Čechové*) and, according to legend, their leader Čech, who brought them to Bohemia, to settle on Říp Mountain. The etymology of the word *Čech* can be traced back to the Proto-Slavic root **čel-*, meaning "member of the people; kinsman", thus making it cognate to the Czech word *člověk* (a person).[16]

The country has been traditionally divided into lands, namely Bohemia proper (Čechy) in the west, Moravia (Morava) in the southeast, and Czech Silesia (Slezsko; the smaller, south-eastern part of historical Silesia, most of which is located within modern Poland) in the northeast. Known officially as the "Crown of the Kingdom of Bohemia" since the 14th century, a number of other names for the country have been used, including the Lands of the Bohemian Crown, Czech/Bohemian lands, Bohemian Crown, and the Lands of the Crown of Saint Wenceslas.

When the country regained its independence after the dissolution of the Austro-Hungarian empire in 1918, the new name of *Czechoslovakia* was coined to reflect the union of the Czech and Slovak nations within the one country.

Following the dissolution of Czechoslovakia at the end of 1992, the Czech part of the former nation found itself without a common single-word name in English. In 1993, the Czech Ministry of Foreign Affairs suggested the name **Czechia** /ˈtʃɛkiə/ (*Česko* Czech pronunciation: [ˈtʃɛsko] in Czech) as an official alternative in all situations other than formal official documents and the full names of government institutions; however, this has not become widespread in English.

History

Prehistory

Archaeologists have found evidence of prehistoric human settlements in the area, dating back to the Neolithic era. In the classical era, from the 3rd century BC Celtic migrations, the Boii and later in the 1st century, Germanic tribes of Marcomanni and Quadi settled there. During the Migration Period around the 5th century, many Germanic tribes moved westwards and southwards out of Central Europe.

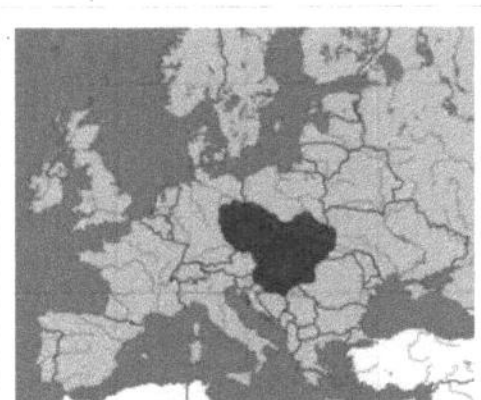

Great Moravia during the reign of Svatopluk I

Czech lion with a double tail from the Small coat of arms first time appeared in the 13th century

Slavic people from the Black Sea-Carpathian region settled in the area (a movement that was also stimulated by the onslaught of peoples from Siberia and Eastern Europe: Huns, Avars, Bulgars and Magyars). In the sixth century they moved southwards into Bohemia, Moravia and some of present day Austria. During the 7th century, the Frankish merchant Samo, supporting the Slavs fighting their Avar rulers, became the ruler of the first known Slav state in Central Europe. The Moravian principality arose in the 8th century and reached its zenith in the 9th, when it held off the influence of the Franks and won the protection of the Pope.

Bohemia

The Bohemian or Czech state emerged in the late 9th century, when it was unified by the Přemyslid dynasty. The Kingdom of Bohemia was, as the only kingdom in the Holy Roman Empire, a significant regional power during the Middle Ages. It was part of the Empire from 1002 till 1806, with the exception of years 1440–1526. In 1212, King Přemysl Otakar I (bearing the title "king" since 1198) extracted a Golden Bull of Sicily (a formal edict) from the emperor, confirming the royal title for Otakar and his descendants and the Duchy of Bohemia was raised to a Kingdom. Czech king should be exempt from all future obligations to the Holy Roman Empire except for participation in the imperial councils. The German immigration occurred in Bohemian periphery in 13th century. The Germans populated towns and mining districts and, in some cases, formed German colonies in the interior of the Czech lands. In 1235, the mighty Mongol army launched an invasion of Europe. After the Battle of Legnica, the Mongols carried their devastating raid into Moravia, but they were beaten by the Czech royal army in a battle of Olomouc, where they killed Genghis Khan grandson Baidar and continued into Hungarian lands.[17]

Přemysl Ottokar II, (c. 1233–1278), King of Bohemia and ruler of Austria, Styria, Carinthia and Carniola

King Přemysl Otakar II earned the nickname "Iron and Golden King" because of his military power and wealth. He acquired Austria, Styria, Carinthia and Carniola, thus spreading the Bohemian territory to the Adriatic Sea. He met his death at the Battle on the Marchfeld in 1278 in a war with his rival, King Rudolph I of Germany.[18] Ottokar's son Wenceslaus II acquired the Polish crown in 1300 for himself and the Hungarian crown for his son. He built a great empire stretching from the Danube river to the Baltic Sea. In 1306, the last king of Přemyslid line was murdered in mysterious circumstances in Olomouc while he was resting. After a series of dynastic wars, the House of Luxembourg gained the Bohemian throne.[19]

Charles IV, 1316–78, eleventh king of Bohemia, elected as the *Největší Čech* (Greatest Czech) of all time.[20]

The 14th century, particularly the reign of Czech King Charles IV (1316–1378), who became also King of Italy, King of the Romans and Holy Roman Emperor, is considered the Golden Age of Czech history. Of particular significance was the founding of Charles University in Prague in 1348, Charles Bridge, Charles Square and were completed much of the Prague Castle and cathedral of Saint Vitus. He acquired the territory of Brandenburg (until 1415), Lusatia (until 1635), and Silesia (until 1742) under Czech crown. The Black Death, which had raged in Europe from 1347 to 1352, decimated the Kingdom of Bohemia in 1380,[21] killing about 10% of the population.[22]

In the 15th century, the religious and social reformer Jan Hus formed a movement later named after him. Although Hus was named a heretic and burnt in Constance in 1415, his followers seceded from the Catholic Church and in the Hussite Wars (1419–1434) defeated five crusades organized against them by the Holy Roman Emperor Sigismund. Petr Chelčický continued with Czech Hussite Reformation movement. During the next two centuries, 90% of the inhabitants became adherents of the Hussite Christian movement.

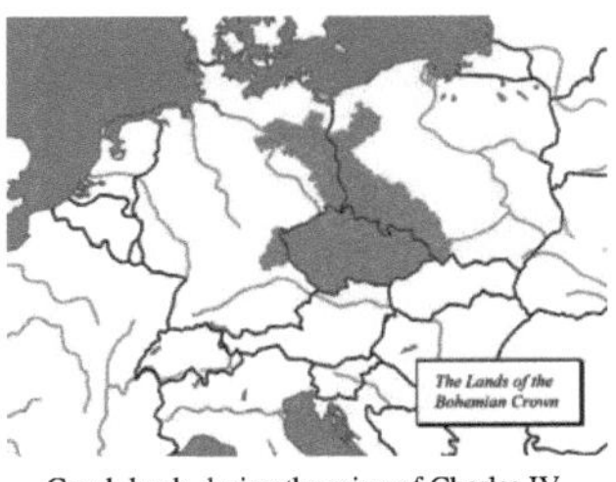

Czech lands during the reign of Charles IV.

After 1526 Bohemia came increasingly under Habsburg control as the Habsburgs became first the elected and then the hereditary rulers of Bohemia. The Defenestration of Prague and subsequent revolt against the Habsburgs in 1618 marked the start of the Thirty Years' War, which quickly spread throughout Central Europe. In 1620, the rebellion in Bohemia was crushed at the Battle of White Mountain, and the ties between Bohemia and Habsburgs' hereditary lands in Austria were strengthened. The war had a devastating effect on the local population; the people were given the choice either to convert to Catholicism or leave the country.

The following period, from 1620 to the late 18th century, has been often called the "Dark Age". The population of the Czech lands declined by a third through war, disease, famine and the expulsion of the Protestant Czechs.[23] The Habsburgs banned all religions other than Catholicism.[24] Ottoman Turks and Tatars invaded Moravia in 1663.[25] In 1679–1680 the Czech lands faced a devastating plague and an uprising of serfs.[26]

The reigns of Maria Theresa of Austria and her son Joseph II, Holy Roman Emperor and co-regent from 1765, were characterized by enlightened absolutism. In 1742, most of Silesia, then the possession of the Bohemian crown, was seized by King Frederick II of Prussia in the War of the Austrian Succession. The Great Famine, which lasted from 1770 until 1771, killed about one tenth of the Czech population, or 250,000 inhabitants, and radicalized countrysides leading to peasant uprisings.[27]

After the fall of the Holy Roman Empire, Bohemia became part of the Austrian Empire and later of Austria–Hungary. Serfdom was not completely abolished until 1848. After the Revolutions of 1848, Emperor Franz Josef I of Austria instituted an absolute monarchy in an effort to balance competing ethnic interests in the empire.

Czechoslovakia

An estimated 1.4 million Czech soldiers fought in World War I, of whom some 150,000 died. More than 90,000 Czech volunteers formed the Czechoslovak Legions in France, Italy and Russia, where they fought against the Central Powers and later against Bolshevik troops.[28] Following the collapse of the Austro-Hungarian Empire after World War I, the independent republic of Czechoslovakia was created in 1918. This new country incorporated the Bohemian Crown

Czechoslovak troops in Vladivostok, Siberia
(1918).

(Bohemia, Moravia and Silesia) and parts of the Kingdom of Hungary (Slovakia and the Carpathian Ruthenia) with significant German, Hungarian, Polish and Ruthenian speaking minorities.[29]

Although Czechoslovakia was a unitary state, it provided what were at the time rather extensive rights to its minorities and remained the only democracy in this part of Europe in the interwar period. The effects of the Great Depression including high unemployment and massive propaganda from Nazi Germany, however, resulted in discontent and strong support among ethnic Germans for a break from Czechoslovakia. Adolf Hitler took advantage of this opportunity and, using Konrad Henlein's separatist Sudeten German Party, gained the largely German speaking Sudetenland (and its substantial Maginot Line-like border fortifications) through the 1938 Munich Agreement (signed by Nazi Germany, France, Britain and Italy), despite the mobilization of 1.2 million-strong Czechoslovak army and the Franco-Czech military alliance. Poland annexed the Zaolzie area around Český Těšín. Hungary gained parts of Slovakia and the Subcarpathian Rus as a result of the First Vienna Award in November 1938.

Tomáš Garrigue Masaryk, first president of Czechoslovakia.

The remainders of Slovakia and the Subcarpathian Rus gained greater autonomy, with the state renamed to "Czecho-Slovakia". After Nazi Germany threatened to annex part of Slovakia, allowing the remaining regions to be partitioned by Hungary and Poland, Slovakia chose to maintain its national and territorial integrity, seceding from Czecho-Slovakia in March 1939, and allying itself, as demanded by Germany, with Hitler's coalition.[30]

The remaining Czech territory was occupied by Germany, which transformed it into the so-called Protectorate of Bohemia and Moravia. The protectorate was proclaimed part of the Third Reich, and the president and prime minister were subordinate to the Nazi Germany's *Reichsprotektor* ("imperial protector"). Subcarpathian Rus declared independence as the Republic of Carpatho-Ukraine on 15 March 1939 but was invaded by Hungary the same day and formally annexed the next day. Approximately 345,000 Czechoslovak citizens, including 277,000 Jews, were killed or executed while hundreds of thousands of others were sent to prisons and concentration camps or used as forced labour. Perhaps two–thirds of the Czech nation was destined either for extermination or removal.[31] A Nazi German concentration camp existed at Terezín, north of Prague.

There was a strong Czech resistance to Nazi occupation, both at home and abroad, most notably with the assassination of Nazi German leader Reinhard Heydrich by Czechoslovakian soldiers Jozef Gabčík and Jan Kubiš in a Prague suburb on 27 May 1942. The Czechoslovak government-in-exile and its army fighting against the Germans were acknowledged by the Allies; Czech/Czechoslovak troops fought from the very beginning of the war in Poland, France, the UK, North Africa, the Middle East and the Soviet Union. The German occupation ended on 9 May 1945, with the arrival of the Soviet and American armies and the Prague uprising. An estimated 140,000 Soviet soldiers died in liberating Czechoslovakia from German rule.[32]

A memorial to 82 Lidice children murdered by the Nazi Germans in Chelmno.

In 1945–1946, almost the entire German minority in Czechoslovakia, about 3 million people, were expelled to Germany and Austria. During this time, thousands of Germans were held in prisons and detention camps or used as forced labour. In the summer of 1945, there were several massacres. The only Germans not expelled were some 250,000 who had been active in the resistance against the Nazi Germans or were considered economically important, though many of these emigrated later. Following a Soviet-organised referendum, the Subcarpathian Rus never returned under Czechoslovak rule but became part of the Ukrainian Soviet Socialist Republic, as the Zakarpattia Oblast in 1946.

Czechoslovakia uneasily tried to play the role of a "bridge" between the West and East. However, the Communist Party of Czechoslovakia rapidly increased in popularity, with a general disillusionment with the West, because of the pre-war Munich Agreement, and a favourable popular attitude towards the Soviet Union, because of the Soviets' role in liberating Czechoslovakia from German rule. In the 1946 elections, the Communists gained 38%[33] of the votes and became the largest party in the Czechoslovak parliament. They formed a coalition government with other parties of the National Front and moved quickly to consolidate power. The decisive step took place in February 1948, during a series of events characterized by Communists as a "revolution" and by anti-Communists as a "takeover", the Communist People's Militias secured control of key locations in Prague, and a new all-Communist government was formed.

For the next 41 years, Czechoslovakia was a Communist state within the Eastern Bloc. This period is characterized by lagging behind the West in almost every aspect of social and economic development. The country's GDP per capita fell from the level of neighboring Austria below that of Greece or Portugal in the 1980s. The Communist government completely nationalized the means of production and established a command economy. The economy grew rapidly during the 1950s but slowed down in the 1960s and 1970s and stagnated in the 1980s. The political climate was highly repressive during the 1950s, including numerous show trials and hundreds of thousands of political prisoners, but became more open and tolerant in the late 1960s, culminating in Alexander Dubček's leadership in the 1968 Prague Spring, which tried to create "socialism with a human face" and perhaps even introduce political pluralism. This was forcibly ended by invasion of all Warsaw Pact member countries with the exception of Romania and Albania on 21 August 1968.

The invasion was followed by a harsh program of "Normalization" in the late 1960s and the 1970s. Until 1989, the political establishment relied on censorship of the opposition. Dissidents published Charter 77 in 1977, and the first of a new wave of protests were seen in 1988. Between 1948 and 1989 more than 250,000 Czechs and Slovaks were sent to prison for "anti-state activities" and over 400,000 emigrated.[34]

Velvet revolution and independence

In November 1989, Czechoslovakia returned to a liberal democracy through the peaceful "Velvet Revolution". However, Slovak national aspirations strengthened and on 1 January 1993, the country peacefully

The Czech Republic became a member of the European Union in 2004, signed the Lisbon Treaty in 2007 and ratified it in 2009 as the last EU member.

split into the independent Czech Republic and Slovakia. Both countries went through economic reforms and privatisations, with the intention of creating a capitalist economy. This process was largely successful; in 2006 the Czech Republic was recognised by the World Bank as a "developed country",[12] and in 2009 the Human Development Index ranked it as a nation of "Very High Human Development".[13]

From 1991, the Czech Republic, originally as part of Czechoslovakia and since 1993 in its own right, has been a member of the Visegrád Group and from 1995, the OECD. The Czech Republic joined NATO on 12 March 1999 and the European Union on 1 May 2004. It held the Presidency of the European Union for the first half of 2009.

Politics

The Czech Republic is a pluralist multi-party parliamentary representative democracy, with the Prime Minister as head of government. The Parliament (*Parlament České republiky*) is bicameral, with the Chamber of Deputies (Czech: *Poslanecká sněmovna*) (200 members) and the Senate (Czech: *Senát*) (81 members).

The President of the Czech Republic is selected by a joint session of the parliament for a five-year term, with no more than two consecutive terms. The president is a formal head of state with limited specific powers, most importantly to return bills to the parliament, nominate constitutional court judges for the Senate's approval and dissolve the parliament under certain special and unusual circumstances. He also appoints the prime minister, as well the other members of the cabinet on a proposal by the prime minister. From 2013 on, the president will be elected by the public, not the parliament.[35]

Václav Klaus, current President of the Czech Republic

The Prime Minister is the head of government and wields considerable powers, including the right to set the agenda for most foreign and domestic policy, mobilize the parliamentary majority and choose government ministers.

The members of the Chamber of Deputies are elected for a four-year term by proportional representation, with a 5% election threshold. There are 14 voting districts, identical to the country's administrative regions. The Chamber of Deputies, the successor to the Czech National Council, has the powers and responsibilities of the now defunct federal parliament of the former Czechoslovakia.

Václav Havel, the first President of the Czech Republic

The members of the Senate are elected in single-seat constituencies by two-round runoff voting for a six-year term, with one-third elected every even year in the autumn. The first election was in 1996, for differing terms. This arrangement is modeled on the U.S. Senate, but each constituency is roughly the same size and the voting system used is a two-round runoff. The Senate is unpopular among the public and suffers from low election turnout, overall roughly 30% in the first round and 20% in the second.

Foreign relations

Membership in the European Union is central in Czech Republic's foreign policy. The Czech Republic held the Presidency of the Council of the European Union for the first half of 2009.

Czech officials have supported dissenters in Burma, Belarus, Moldova and Cuba.[36]

Czech soldier in Afghanistan

Military

The Czech armed forces consist of the Army, Air Force and of specialized support units. The president of the Czech Republic, is Commander-in-Chief of the armed forces. In 2004 the army transformed itself into a fully professional organisation and compulsory military service was abolished. The country has been a member of NATO since 12 March 1999. Defence spending is around 1.8% of the GDP (2006). Currently, as a member of NATO, the Czech military are participating in ISAF and KFOR operations and have soldiers in Afghanistan and Kosovo. Main equipment includes: multirole fighters JAS-39 Gripen, combat aircraft Aero L-159 Alca, attack helicopters Mi-24, armoured vehicles Pandur II, OT-64, OT-90, BVP-2 and Czech modernised tanks T-72 (T-72M4CZ).

Administrative divisions

Since 2000, the Czech Republic is divided into thirteen regions (Czech: *kraje*, singular *kraj*) and the capital city of Prague. Each region has its own elected Regional Assembly (*krajské zastupitelstvo*) and *hejtman* (usually translated as hetman or "president"). In Prague, their powers are executed by the city council and the mayor.

The older seventy-six districts (*okresy*, singular *okres*) including three "statutory cities" (without Prague, which had special status) lost most of their importance in 1999 in an administrative reform; they remain as territorial divisions and seats of various branches of state administration.[37]

Map of the Czech Republic with traditional regions and current administrative regions

Map with districts.

(Lic. plate)	Region	Administrative seat	Population (2004 est.)	Population (2010 est.)
A	Prague, the Capital City (*Hlavní město Praha*)		1,170,571	1,251,072
S	Central Bohemian Region (*Středočeský kraj*)	offices located in Prague (Praha)	1,144,071	1,256,850
C	South Bohemian Region (*Jihočeský kraj*)	České Budějovice	625,712	637,723
P	Plzeň Region (*Plzeňský kraj*)	Plzeň	549,618	571,831
K	Karlovy Vary Region (*Karlovarský kraj*)	Karlovy Vary	304,588	307,380
U	Ústí nad Labem Region (*Ústecký kraj*)	Ústí nad Labem	822,133	835,814
L	Liberec Region (*Liberecký kraj*)	Liberec	427,563	439,458
H	Hradec Králové Region (*Královéhradecký kraj*)	Hradec Králové	547,296	554,370
E	Pardubice Region (*Pardubický kraj*)	Pardubice	505,285	516,777
M	Olomouc Region (*Olomoucký kraj*)	Olomouc	635,126	641,555
T	Moravian-Silesian Region (*Moravskoslezský kraj*)	Ostrava	1,257,554	1,244,837
B	South Moravian Region (*Jihomoravský kraj*)	Brno	1,123,201	1,152,819
Z	Zlín Region (*Zlínský kraj*)	Zlín	590,706	590,527
J	Vysočina Region (*Kraj Vysočina*)	Jihlava	517,153	514,805

Geography

The Czech Republic lies mostly between latitudes 48° and 51° N (a small area lies north of 51°), and longitudes 12° and 19° E.

The Czech landscape is exceedingly varied. Bohemia, to the west, consists of a basin drained by the Elbe (Czech: *Labe*) and the Vltava (or Moldau) rivers, surrounded by mostly low mountains, such as the Krkonoše range of the Sudetes. The highest point in the country, Sněžka at 1602 m (5256 ft), is located here. Moravia, the eastern part of the country, is also quite hilly. It is drained mainly by the Morava River, but it also contains the source of the Oder River (Czech: *Odra*).

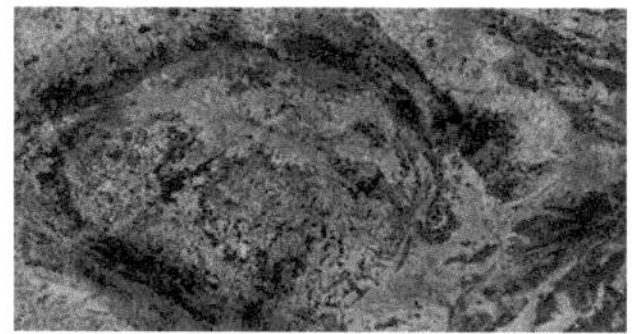
Satellite image of the Czech Republic

Water from the landlocked Czech Republic flows to three different seas: the North Sea, Baltic Sea and Black Sea. The Czech Republic also leases the Moldauhafen, a 30000-square-metre (**unknown operator: u'strong'**-acre) lot in the middle of the Hamburg Docks, which was awarded to Czechoslovakia by Article 363 of the Treaty of Versailles, to allow the landlocked country a place where goods transported down river could be transferred to seagoing ships. The territory reverts to Germany in 2028.

Czech Switzerland is one of four Czech national parks

Phytogeographically, the Czech Republic belongs to the Central European province of the Circumboreal Region, within the Boreal Kingdom. According to the World Wide Fund for Nature, the territory of the Czech Republic can be subdivided into four ecoregions: the Central European mixed forests, Pannonian mixed forests, Western European broadleaf forests and Carpathian montane conifer forests.

There are four national parks in the Czech Republic. The oldest is Krkonoše National Park (Biosphere Reserve), Šumava National Park (Biosphere Reserve), Podyjí National Park, Bohemian Switzerland.

Climate

Šance Dam, part of the Moravian-Silesian Beskids

The Czech Republic has a temperate continental climate, with relatively hot summers and cold, cloudy and snowy winters. The temperature difference between summer and winter is relatively high, due to the landlocked geographical position.[38]

Within the Czech Republic, temperatures vary greatly, depending on the elevation. In general, at higher altitudes, the temperatures decrease and precipitation increases. The wettest area in the Czech Republic is found around Bílý Potok in Jizera Mountains and the driest region is the Louny District to the northwest of Prague. Another important factor is the distribution of the mountains; therefore, the climate is quite varied.

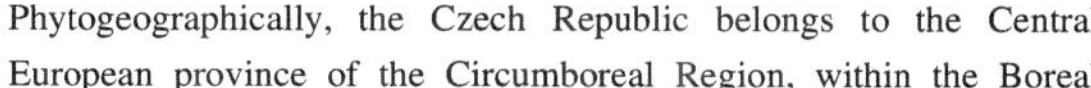

At the highest peak of Sněžka (1602 m/5256 ft), the average temperature is only −0.4 °C (31 °F), whereas in the lowlands of the South Moravian Region, the average temperature is as high as 10 °C (**unknown operator: u'strong' °F**). The country's capital, Prague, has a similar average temperature, although this is influenced by urban factors.

The coldest month is usually January, followed by February and December. During these months, there is usually snow in the mountains and sometimes in the major cities and lowlands. During

Rolling hills of Králický Sněžník

March, April and May, the temperature usually increases rapidly, especially during April, when the temperature and weather tends to vary widely during the day. Spring is also characterized by high water levels in the rivers, due to melting snow with occasional flooding.

The warmest month of the year is July, followed by August and June. On average, summer temperatures are about 20 °C higher than during winter. Temperatures above 30 °C (**unknown operator: u'strong' °F**) are not unusual. Summer is also characterized by rain and storms.

Autumn generally begins in September, which is still relatively warm and dry. During October, temperatures usually fall below 15 °C (**unknown operator: u'strong' °F**) or 10 °C (**unknown operator: u'strong' °F**) and deciduous trees begin to shed their leaves. By the end of November, temperatures usually range around the freezing point.

The coldest temperature ever measured was in Litvínovice near České Budějovice in 1929, at −42.2 °C (−**unknown operator: u'strong' °F**) and the hottest measured, was at 40.4 °C (**unknown operator: u'strong' °F**) in Dobřichovice in 2012.[39]

Dukovany Nuclear Power Station

Most rain falls during the summer. Sporadic rainfall is relatively constant throughout the year (in Prague, the average number of days per month experiencing at least 0.1 mm of rain varies from 12 in September and October to 16 in November) but concentrated heavy rainfall (days with more than 10 mm per day) are more frequent in the months of May to August (average around two such days per month).[40]

Economy

The Czech Republic possesses a developed,[41] high-income[42] economy with a GDP per capita of 80% of the European Union average.[43] One of the most stable and prosperous of the post-Communist states, the Czech Republic saw growth of over 6% annually in the three years before the outbreak of the recent global economic crisis. Growth has been led by exports to the European Union, especially Germany, and foreign investment, while domestic demand is reviving.

Most of the economy has been privatised, including the banks and telecommunications. The current centre-right government plans to continue with privatisation, including the energy industry and the Prague airport. It has recently agreed to the sale of a 7% stake in the energy producer, CEZ Group, with the sale of the Budějovický Budvar brewery also mooted. A 2009 survey in cooperation with the Czech Economic Association found that the majority of Czech economists favor continued liberalization in most sectors of the economy.[44]

The country is part of the Schengen Area from 1 May 2004, having abolished border controls, completely opening its borders with all of its neighbours, Germany, Austria, Poland and Slovakia, on 21 December 2007.[45] The Czech Republic became a member of the World Trade Organisation.

The last Czech government led by social democrats had expressed a desire to adopt the euro in 2010, but the current centre-right government suspended that plan in 2007.[46] An exact date has not been set up, but the Finance Ministry described adoption by 2012 as realistic,[47] if public finance reform passes. However, the most recent draft of the euro adoption plan omits giving any date. Although the country is economically better positioned than other EU Members to adopt the euro, the change is not expected before 2019, due to political reluctance on the matter.[48]

The Programme for International Student Assessment, coordinated by the OECD, currently ranks the Czech education system as the 15th best in the world, higher than the OECD average.[49] The Czech Republic is ranked 30th in the 2012 Index of Economic Freedom.

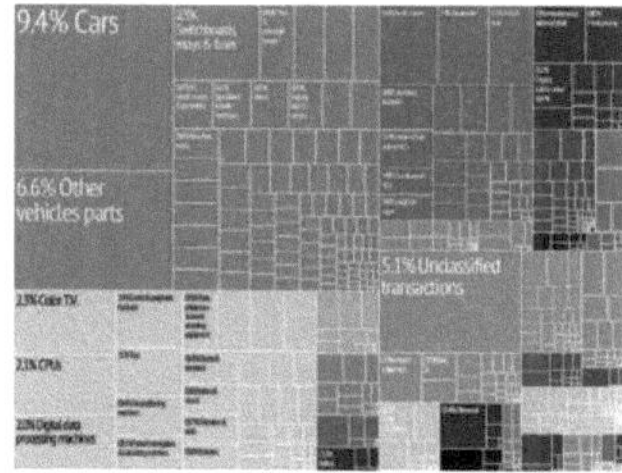

Graphical depiction of The Czech Republic's product exports in 28 color coded categories.

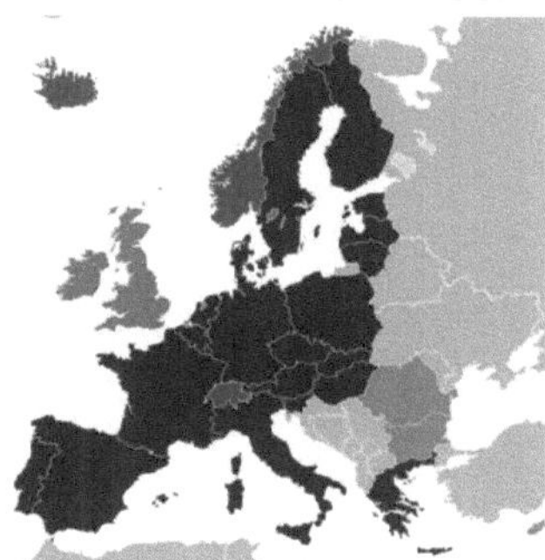

The Czech Republic is part of the EU single market and the Schengen Area.

Headquarters of Czech National Bank in Prague

Škoda Auto is one of the largest car manufacturers in Central Europe. In 2011, sold a record number of 875,000 cars and said it aimed to double sales by 2018

Transportation infrastructure

Double deck trains called CityElefant made by Škoda Works operate near larger cities

Václav Havel Airport Prague

Václav Havel Airport in Prague is the main international airport in the country. In 2010, it handled 11.6 million passengers, which makes it the busiest airport in Central and Eastern Europe. In total, Czech Republic has 46 airports with paved runways, six of which provide international air services in Brno, Karlovy Vary, Mošnov (near Ostrava), Pardubice, Prague and Kunovice (near Uherské Hradiště).

Expressway R1 bypassing Prague

České dráhy (the Czech railways) is the main railway operator in the Czech Republic, with about 180 million passengers carried yearly. Its cargo division, ČD Cargo, is the fifth largest railway cargo operator in the European Union. With 9505 km (5906.13 mi) of tracks, the Czech Republic has one of the densest railway networks in Europe.[50] Of that number, 2926 km (1818.13 mi) is electrified, 7617 km (4732.98 mi) are single-line tracks and 1866 km (1159.48 mi) are double and multiple-line tracks.[51] In 2006 the new Italian tilting trains Pendolino ČD Class 680 entered service. They have reached a speed of 237 km/h setting a new Czech railway speed record.

In 2005, according to the Czech Statistical Office, 65.4 percent of electricity was produced by steam, combined and combustion power plants (mostly coal); 30 percent in nuclear plants; and 4.6 percent from renewable sources, including hydropower. Russia, via pipelines through Ukraine and to a lesser extent, Norway, via pipelines through Germany, supply the Czech Republic with liquid and natural gas.

The Czech Republic is reducing its dependence on highly polluting low-grade brown coal as a source of energy. Nuclear power presently provides about 30 percent of the total power needs, its share is projected to increase to 40 percent. Natural gas is procured from Russian Gazprom, roughly three-fourths of domestic consumption and from Norwegian companies, which make up most of the remaining one-fourth. Russian gas is imported via Ukraine (Druzhba pipeline), Norwegian gas is transported through Germany. Gas consumption (approx. 100 TWh in 2003–2005) is almost double electricity consumption. South Moravia has small oil and gas deposits.

Electric multiple unit Pendolino at Prague main railway station

The road network in the Czech Republic is 55653 km (34581.17 mi) long.[52] and 738,4 km of motorways and 439,1 km of expressways.[53] The speed limit is 50 km/h within towns, 90 km/h outside of towns and 130 km/h on expressways.

Communications

The Czech Republic ranks in the top 10 countries worldwide with the fastest average internet speed.[54] The Czech Republic has the most Wi-Fi subscribers in the European Union.[55] [56] By the beginning of 2008, there were over 800 mostly local WISPs,[57] [58] with about 350,000 subscribers in 2007. Plans based on either GPRS, EDGE, UMTS or CDMA2000 are being offered by all three mobile phone operators (T-Mobile, Vodafone, Telefónica O2) and internet provider U:fon. Government-owned Český Telecom slowed down broadband penetration. At the beginning of 2004, local-loop unbundling began and alternative operators started to offer ADSL and also SDSL. This and later privatisation of Český Telecom helped drive down prices.

On 1 July 2006, Český Telecom was acquired by globalized company (Spain owned) Telefónica group and adopted new name Telefónica O2 Czech Republic. As of April 2012, VDSL and ADSL2+ are offered in many variants, without with speeds up to 25 Mbit/s. Cable internet is gaining popularity with its higher download speeds beginning at 2 Mbit/s up to 120 Mbit/s.

Science

The Czech Republic has a rich scientific tradition. Important inventions include the modern contact lens, the separation of modern blood types, and the production of the Semtex plastic explosive. Prominent scientists who lived and worked in historically Czech lands include:

The modern contact lens was invented by Otto Wichterle and Drahoslav Lím.

- John Amos Comenius (1592–1670), teacher, educator and the founder of modern education.[59]
- Václav Prokop Diviš (1698–1765), inventor of the first grounded lightning rod.
- Bernard Bolzano (1781–1848), noted mathematician, logician, philosopher, and pacifist.
- Jan Evangelista Purkyně (1787–1869), anatomist and physiologist responsible for the discovery of Purkinje cells, Purkinje fibres and sweat glands, as well as Purkinje images and the Purkinje shift.
- Josef Ressel (1793–1857), inventor of the screw propeller.[59]
- Gregor Mendel (1822–1884), often called the "*father of genetics*", is famed for his research concerning the inheritance of genetic traits.[59]
- Jakub Husník (1837–1916), inventor of the improved photolithography.
- Karel Klíč (1841–1926), painter and photographer, inventor of the photogravure.
- František Křižík (1847–1941), electrical engineer, inventor of the arc lamp.
- Jan Janský (1873–1921), serologist and neurologist, discovered classification of blood into the four types.
- Bedřich Hrozný (1879–1952), deciphered the Hittite language.[59]
- Jaroslav Heyrovský (1890–1967), inventor of the polarography, electroanalytical chemistry and recipient of the Nobel Prize.[59]
- Otto Wichterle (1913–1998) and Drahoslav Lím (1925–2003), Czech chemists responsible for the invention of the modern contact lens.[60]
- Stanislav Brebera (* 1925), inventor of the plastic explosive Semtex.[61]
- Antonín Holý (1936–2012), scientist and chemist, in 2009 was involved in the creation of the most effective drug in the treatment of AIDS.[62]
- Josef Čapek (1887–1945), painter, writer and cartoonist, and Karel Čapek (1890–1938), also a writer; brothers who originated the word robot in the play R.U.R. (Rossum's Universal Robots).

A number of other scientists are also connected in some way with the Czech Lands, including astronomers Johannes Kepler and Tycho Brahe, the founder of the psychoanalytic school of psychiatry Sigmund Freud, physicists Ernst Mach, Nikola Tesla, Albert Einstein, engineer Viktor Kaplan and logician Kurt Gödel.

Tourism

The Czech economy gets a substantial income from tourism. In 2011, Prague was the sixth most visited city in Europe.[63] In 2001, the total earnings from tourism reached 118.13 billion CZK, making up 5.5% of GNP and 9.3% of overall export earnings. The industry employs more than 110,000 people – over 1% of the population.[64] In 2008, however, there was a slump in tourist numbers in Prague, possibly due to the strong Czech koruna (crown) making the country too expensive for visitors, compared to the level of services that were available.[65]

Since the fall of the Iron Curtain in 1989, Prague has become one of the most visited cities in Europe

The country's reputation has also suffered with guidebooks and tourists reporting overcharging by taxi drivers and pickpocketing problems.[65] [66] Since 2005, Prague's mayor, Pavel Bém, has worked to improve this reputation by cracking down on petty crime[66] and, aside from these problems, Prague is a safe city.[67] Also, the Czech Republic as a whole generally has a low crime rate.[68] For tourists, the Czech Republic is considered a safe destination to visit. The low crime rate makes most cities and towns safe to walk around even after dark.

There are several centres of tourist activity. The spa towns, such as Karlovy Vary, Mariánské Lázně and Františkovy Lázně, are particularly popular holiday destinations. Other popular tourist sites are the many castles and chateaux, such as those at Karlštejn Castle, Český Krumlov and the Lednice–Valtice area. Away from the towns, areas such as Český ráj, Šumava and the Krkonoše Mountains attract visitors seeking outdoor pursuits.

The country is also famous for its love of puppetry and marionettes with a number of puppet festivals throughout the country.

Aquapalace Praha in Čestlice near Prague, is the biggest water park in central Europe.[69]

The Czech Republic also has a number of beer festivals, including: Czech Beer Festival (the biggest Czech beer festival, it is 17 days long and held every year in May in Prague), Pilsner Fest (every year in August in Plzeň), The "Olomoucký pivní festival" (in Olomouc) or festival "Slavnosti piva v Českých Budějovicích" (in České Budějovice).

Demographics

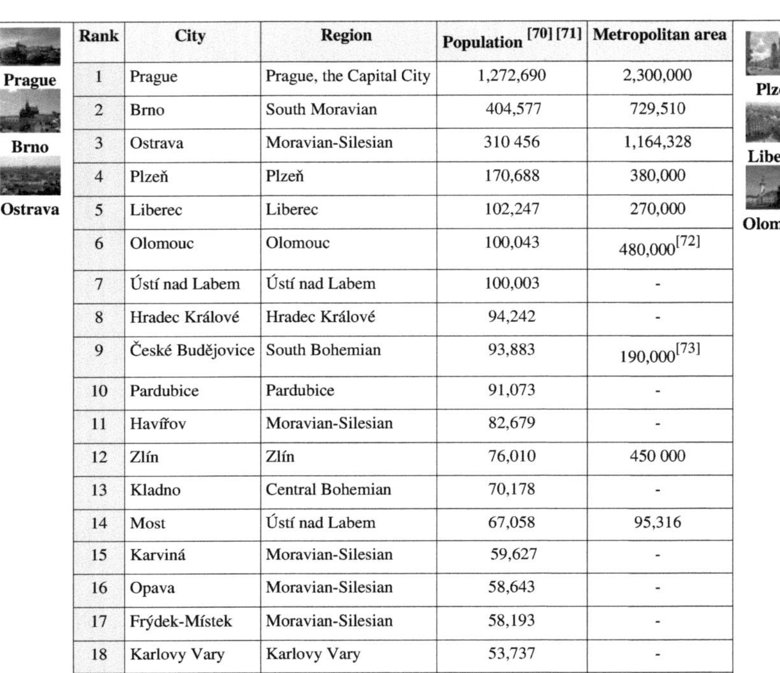

	Rank	City	Region	Population [70] [71]	Metropolitan area	
	1	Prague	Prague, the Capital City	1,272,690	2,300,000	
	2	Brno	South Moravian	404,577	729,510	
	3	Ostrava	Moravian-Silesian	310 456	1,164,328	
	4	Plzeň	Plzeň	170,688	380,000	
	5	Liberec	Liberec	102,247	270,000	
	6	Olomouc	Olomouc	100,043	480,000[72]	
	7	Ústí nad Labem	Ústí nad Labem	100,003	-	
	8	Hradec Králové	Hradec Králové	94,242	-	
	9	České Budějovice	South Bohemian	93,883	190,000[73]	
	10	Pardubice	Pardubice	91,073	-	
	11	Havířov	Moravian-Silesian	82,679	-	
	12	Zlín	Zlín	76,010	450 000	
	13	Kladno	Central Bohemian	70,178	-	
	14	Most	Ústí nad Labem	67,058	95,316	
	15	Karviná	Moravian-Silesian	59,627	-	
	16	Opava	Moravian-Silesian	58,643	-	
	17	Frýdek-Místek	Moravian-Silesian	58,193	-	
	18	Karlovy Vary	Karlovy Vary	53,737	-	
	19	Jihlava	Vysočina	50,760	-	
	20	Děčín	Ústí nad Labem	50,620	-	

Historical population

Year	Pop.	±%
1857	7016531	—
1869	7617230	+8.6%
1880	8222013	+7.9%
1890	8665421	+5.4%
1900	9372214	+8.2%
1910	10078637	+7.5%
1921	10009587	−0.7%
1930	10674386	+6.6%
1950	8896133	−16.7%
1961	9571531	+7.6%
1970	9807697	+2.5%
1980	10291927	+4.9%
1991	10302215	+0.1%
2001	10230060	−0.7%
2011	10436560	+2.0%

According to preliminary results of the 2011 census, the majority of the inhabitants of the Czech Republic are Czechs (63.7%), followed by Moravians (4.9%), Slovaks (1.4%), Poles (0.4%), Germans (0.2%) and Silesians (0.1%). As the 'nationality' was an optional item, a substantial number of people left this field blank (26.0%).[74] According to some estimates, there are about 250,000 Romani people in the Czech Republic.[75] [76]

There were 436,116 foreigners residing in the country in October 2009, according to the Czech Interior Ministry,[77] with the largest groups being Ukrainian (132,481), Slovak (75,210), Vietnamese (61,102), Russian (29,976), Polish (19,790), German (14,156), Moldovan (10,315), Bulgarian (6,346), Mongolian (5,924), American (5,803), Chinese (5,314), British (4,461), Belarusian (4,441), Serbian (4,098), Romanian (4,021), Kazakh (3,896), Austrian (3,114), Italian (2,580), Dutch (2,553), French (2,356), Croatian (2,351), Bosnian (2,240), Armenian (2,021), Uzbek (1,969), Macedonian (1,787) and Japanese (1,581).[77]

A footpath on Wenceslas Square

The Jewish population of Bohemia and Moravia, 118,000 according to the 1930 census, was virtually annihilated by the Nazi Germans during the Holocaust.[78] There were approximately 4,000 Jews in the Czech Republic in 2005.[79] The former Czech prime minister, Jan Fischer, is of Jewish origin and faith.[80]

Estimates of Czech fertility rate in 2012 are among the lowest in the world at 1.27 children per woman.[81] Immigration increased the population by almost 1% in 2007. About 77,000 new foreigners settle down in the Czech Republic every year.[82] Vietnamese immigrants began settling in the Czech Republic during the Communist period, when they were invited as guest workers by the Czechoslovak government.[83] In 2009, there were about 70,000 Vietnamese in the Czech Republic.[84] In contrast to Ukrainians, Vietnamese come to the Czech Republic to live permanently.[85]

At the turn of the 20th century, Chicago was the city with the third largest Czech population,[86] after Prague and Vienna.[87] According to the 2006 US census, there are 1,637,218 Americans of full or partial Czech descent.[88]

Religion

Religion in the Czech Republic (2011)[89]

Undeclared (45.2%)

Irreligion (34.2%)

Protestantism (0.8%)

Other religions (9.4%)

Roman Catholicism (10.4%)

The Czech Republic has one of the least religious populations in the world. Historically, the Czech people have been characterised as "tolerant and even indifferent towards religion".[90] According to the 2011 census, 34.2% of the population stated they had no religion, 10.3% was Roman Catholic, 0.8% was Protestant (0.5% Czech Brethren and 0.4% Hussite), and 9.4% followed other forms of religion both denominational or not (of which 863 people answered they are Pagan) 45.2% of the population did not answer the question about religion.[89] From 1991 to 2001 and further to 2011 the adherence to Roman Catholicism decreased from 39.0 to 26.8 and then to 10.3; Protestantism similarly declined from 3.7% to 2.1% and then to 0.8%.[91]

According to a Eurobarometer Poll in 2005,[92] 19% of Czech citizens responded that "they believe there is a God" (the second lowest rate among European Union countries after Estonia with 16%),[93] whereas 50% answered that "they believe there is some sort of spirit or life force" and 30% said that "they do not believe there is any sort of spirit, God or life force".

Culture

Music

Music in the Czech Republic has its roots in more than 1,000 year old sacred music (the first surviving references come from the end of the 10th century), in the traditional folk music of Bohemia, Moravia and Silesia and in the long-term high-culture classical music tradition. Since the early eras of artificial music, Czech musicians and composers have often been influenced by genuine folk music. Notable Czech composers include Antonín Dvořák, Bedřich Smetana, Gustav Mahler (he was born and grew up in Czech republic), Adam Michna, Jan Dismas Zelenka, Josef Mysliveček, Leoš Janáček, Josef Suk, Bohuslav Martinů, Erwin Schulhoff and Petr Eben. The most famous music festival is "the Prague Spring" (Pražské jaro), that has been organized annually since 1946.

The Statue of famous composer Antonín Dvořák in Prague

Literature

Czech literature is the literature written by Czechs or other inhabitants of the Czech state, mostly in the Czech language, although other languages like Old Church Slavonic, Latin or German have been also used, especially in the past. Czech authors who had written in the German language, such as Franz Kafka, are usually excluded from the corpus of Czech literature, regardless of their own national self-identification.

The Slav Epic from Alphonse Mucha (1912)

Czech literature is divided into several main time periods: the Middle Ages; the Hussite period; the years of re-Catholicization and the baroque; the Enlightenment and Czech reawakening in the 19th century; the avantgarde of the interwar period; the years under Communism and the Prague Spring; and the literature of the post-Communist Czech Republic. Czech literature and culture played a major role on at least two occasions, when Czechs lived under oppression and political activity was suppressed. On both of these occasions, in the early 19th century and then again in the 1960s, the Czechs used their cultural and literary effort to strive for political freedom, establishing a confident, politically aware nation.

Theatre

Theatre of the Czech Republic has rich tradition with roots in the Middle Ages. In the 19th century, the theatre played an important role in the national awakening movement and later, in the 20th century it became a part of the modern European theatre art.

Film

The Czech Republic has many popular film locations.[94] Filmmakers have come to Prague to shoot scenery no longer found in Berlin, Paris and Vienna. The city of Karlovy Vary was used as a location for the 2006 James Bond film Casino Royale.[95]

Art

The Czech Republic is known worldwide for their individually made, mouth blown and decorated art glass and crystal. One of the best Czech painter and decorative artist was Alphonse Mucha (1860–1939) mainly known for art nouveau posters and his cycle of 20 large canvases named The Slav Epic, which depicts the history of Czechs and other Slavic peoples. The Slav Epic can be now seen in Veletržní Palace in Prague until 2013.

Cuisine

Pilsner Urquell was the first "pilsner" type beer in the world and also the first lager

Czech cuisine is marked by a strong emphasis on meat dishes. Pork is quite common; beef and chicken are also popular. Goose, duck, rabbit and wild game are served. Fish is rare, with the occasional exception of fresh trout and carp, which is served at Christmas.

Czech beer has a long and important history. The first brewery is known to have existed in 1118 and the Czech Republic has the highest beer consumption per capita in the world. The famous Pilsener style beer originated in the western Bohemian city of Plzeň, and further south the town of České Budějovice, known as Budweis in German, lent its name to its beer, eventually known as Budweiser Budvar. Apart from these and other major brands, the Czech Republic also boasts a growing number of top quality small breweries and mini-breweries seeking to continue the age-old tradition of quality and taste, whose output matches the best in the world: Štiřín, Chýně, Oslavany, Kácov. Tourism is slowly growing around the Southern Moravian region too, which has been producing wine since the Middle Ages; about 94% of vineyards in the Czech Republic are Moravian. Aside from Slivovitz, Czech beer and wine, the Czechs also produce two unique liquors, Fernet Stock and Becherovka. Kofola is a non-alcoholic domestic cola soft drink which competes with Coca Cola and Pepsi in popularity.

Svíčková na smetaně is a signature Czech dish, consisting of marinated beef with Czech dumplings (knedlíky)

Unique Czech dishes include roast pork with bread dumplings and stewed cabbage *Vepřo-knedlo-zelo*, roast sirloin beef with steamed dumplings and cream-of-vegetable sauce *Svíčková na smetaně*, tomato sauce *Rajská* or dill sauce *Koprovka*, roast duck with bread or potato dumplings and braised red cabbage, a variety of beef and pork goulash stews *Guláš*, fried cheese *Smažák* or the famous potato pancakes *Bramboráky*, besides a large variety of delicate local sausages, wurst, pâtés and smoked meats and other traditional local foods. Czech desserts include a wide variety of whipped cream, chocolate and fruit pastries and tarts, crepes, creme desserts and cheese, poppy seed filled and other types of traditional cakes *buchty* and Kolache.

Sports

Sports play a part in the life of many Czechs, who are generally loyal supporters of their favorite teams or individuals. The three leading sports in the Czech Republic are ice hockey, football and sport shooting,[96] with the first two drawing the largest attention of both the media and supporters. Tennis is also a very popular sport in the Czech Republic. The many other sports with professional leagues and structures include basketball, volleyball, team handball, track and field athletics and floorball. The Czech ice hockey team won the gold medal at the 1998 Winter Olympics and has won six gold medals at the World Championships including three straight from 1999 to 2001. In total the country has won 10 gold medals in summer (plus 49 as Czechoslovakia) and five gold medals (plus two as Czechoslovakia) in winter Olympic history.

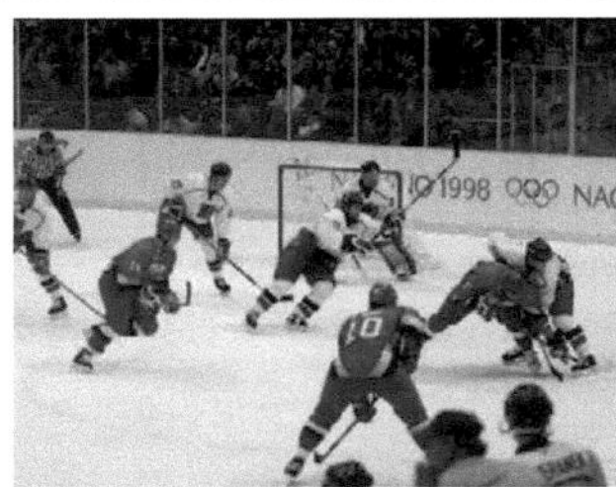

Team Czech Republic beat Russia 1:0 at the 1998 Winter Olympics in Nagano and won gold medal.

Sport is a source of strong waves of patriotism, usually rising several days or weeks before an event. The events considered the most important by Czech fans are: the Ice Hockey World Championships, Olympic Ice hockey tournament, UEFA European Football Championship, FIFA World Cup and qualification matches for such events. In general, any international match of the Czech ice hockey or football national team draws attention, especially when played against a traditional rival. The Czech Republic also has great influence on tennis with such players as, Ivan Lendl, 8 times Grand Slam singles champion, 2010 Wimbledon Championships – Men's Singles finalist Tomáš Berdych, 2011 Wimbledon Championships – Women's Singles champion, Petra Kvitova, 1998 Wimbledon Women's Singles title Jana Novotná, 2011 Wimbledon Championships – Women's Doubles champion Květa Peschke and 18 time Grand Slam Champion Martina Navratilova.

Prague Astronomical Clock is the oldest working astronomical clock in the world	Karlštejn Castle in the Central Bohemian Region, founded in 1348 by Charles IV	Jaromír Jágr is the leading point scorer among active NHL players[97]	Mariánské Lázně, a spa town in the Karlovy Vary Region	Český Krumlov Castle in the South Bohemian Region	Royal castle in Hluboká nad Vltavou, built in the 13th century
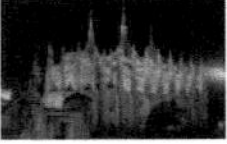					
St. Barbara Church in Kutná Hora	The folk costume of kroje, seen in Vlčnov, Moravia	Český Šternberk is an early gothic castle from the mid 13th century	The Ski resort in the Krkonoše mountains	Chateau in Lednice, a village containing a palace and the largest park in the country	The Czech Crown Jewels are the fourth oldest in Europe

See also

- List of Czech Republic-related topics
- Outline of the Czech Republic

References

[1] "Czech language" (http://www.czech.cz/en/67019-czech-language). *Czech Republic – Official website*. Ministry of Foreign Affairs of the Czech Republic. . Retrieved 14 November 2011.

[2] Citizens belonging to minorities, which traditionally and on long-term basis live within the territory of the Czech Republic, enjoy the right to use their language in communication with authorities and in front of the courts of law (for the list of recognized minorities see National Minorities Policy of the Government of the Czech Republic (http://www.vlada.cz/en/pracovni-a-poradni-organy-vlady/rnm/ historie-a-soucasnost-rady-en-16666/)). The article 25 of the Czech Charter of Fundamental Rights and Basic Freedoms ensures right of the national and ethnic minorities for education and communication with authorities in their own language. Act No. 500/2004 Coll. (*The Administrative Rule*) in its paragraph 16 (4) (*Procedural Language*) ensures, that a citizen of the Czech Republic, who belongs to a national or an ethnic minority, which traditionally and on long-term basis lives within the territory of the Czech Republic, have right to address an administrative agency and proceed before it in the language of the minority. In case that the administrative agency doesn't have an employee with knowledge of the language, the agency is bound to obtain a translator at the agency's own expense. According to Act No. 273/2001 (*About The Rights of Members of Minorities*) paragraph 9 (*The right to use language of a national minority in dealing with authorities and in front of the courts of law*) the same applies for the members of national minorities also in front of the courts of law.

[3] Slovak language may be considered an official language in the Czech Republic under certain circumstances, which is defined by several laws – e.g. law 500/2004, 337/1992. Source: http://portal.gov.cz.Cited: "Například Správní řád (zákon č. 500/2004 Sb.) stanovuje: "V řízení se jedná a písemnosti se vyhotovují v českém jazyce. Účastníci řízení mohou jednat a písemnosti mohou být předkládány i v jazyce slovenském..." (§16, odstavec 1). Zákon o správě daní a poplatků (337/1992 Sb.) „Úřední jazyk: Před správcem daně se jedná v jazyce českém nebo slovenském. Veškerá písemná podání se předkládají v češtině nebo slovenštině..." (§ 3, odstavec 1). http://portal.gov.cz

[4] **(Czech)** ČSÚ – Czech Statistical Office (http://www.czso.cz/sldb2011/redakce.nsf/i/ predbezne_vysledky_scitani_lidu_domu_a_bytu_2011). Czso.cz. Retrieved on 12 August 2012.

[5] "Population" (http://www.czso.cz/eng/redakce.nsf/i/population). Czech Statistical Office. 2012-10-19. . Retrieved 2012-12-19.

[6] Census of Population and Housing 2011: Basic final results (http://www.scitani.cz/sldb2011/eng/redakce.nsf/i/ census_2011_basic_final_results/$File/SLDB-ZAKL-CR_en.xls). Czech Statistical Office (http://www.czso.cz/eng/redakce.nsf/i/ home). Retrieved on 19 December 2012.

[7] "Czech Republic" (http://www.imf.org/external/pubs/ft/weo/2012/02/weodata/weorept.aspx?pr.x=24&pr.y=17&sy=2012& ey=2012&scsm=1&ssd=1&sort=country&ds=.&br=1&c=935,939,936,961&s=NGDP,NGDPD,NGDPPC,NGDPDPC,PPPGDP,PPPPC& grp=0&a=). International Monetary Fund. . Retrieved 10 October 2012.

[8] "Human Development Report 2011" (http://hdrstats.undp.org/en/countries/profiles/CZE.html). United Nations. 2011. . Retrieved 5 November 2011.

[9] Oxford English Dictionary (http://www.oed.com/), second edition, Oxford University Press, 1989.

[10] Mlsna, Petr; Šlehofer F. and Urban D. (2010). "The Path Of Czech Constitutionality" (http://www.vlada.cz/assets/udalosti/vystavy/ Cesty-ceske-ustavnosti.pdf) (in : (Bilingual) - Czech, English). *1st edition*. Praha: Úřad Vlády České Republiky (The Office of the Government of the Czech Republic). pp. 10–11. . Retrieved 31 October 2012.

[11] Čumlivski, Denko (2012). "800 let Zlaté buly sicilské" (http://www.nacr.cz/zpravy/zlatabula800.aspx) (in czech). National Archives of the Czech Republic (Národní Archiv České Republiky). Archived from on 1 October 2012. . Retrieved 31 October 2012.

[12] Velinger, Jan (28 February 2006). "World Bank Marks Czech Republic's Graduation to 'Developed' Status" (http://www.radio.cz/en/ article/76314). Radio Prague. . Retrieved 22 January 2007.

[13] "Human Development Report 2009" (http://hdr.undp.org/en/media/HDR_2009_EN_Complete.pdf) (PDF). *UNDP.org*. . Retrieved 25 April 2010.

[14] "Oxford English Dictionary" (http://oxforddictionaries.com/view/entry/m_en_gb0201690#m_en_gb0201690). Askoxford.com. . Retrieved 4 March 2011.

[15] Czech (http://www.collinsdictionary.com/dictionary/english/czech). CollinsDictionary.com. Collins English Dictionary - Complete & Unabridged 11th Edition. Retrieved November 19, 2012.

[16] Spal, Jaromír. "Původ jména Čech" (http://nase-rec.ujc.cas.cz/archiv.php?art=4320). Naše řeč. . Retrieved 10 December 2012.

[17] Jan Dugosz, Maurice Michael (1997) *The Annals of Jan Dlugosz*, IM Publications, ISBN 1901019004

[18] "The rise and fall of the Przemyslid Dynasty" (http://archiv.radio.cz/history/history03.html). Archiv.radio.cz. . Retrieved 25 April 2010.

[19] "Václav II. český král" (http://www.panovnici.cz/vaclav-II-kral). *panovnici.cz*. .

[20] Emperor Charles IV elected Greatest Czech of all time (http://www.radio.cz/en/article/67495), Radio Prague

[21] "The flowering and the decline of the Czech medieval state" (http://www.arts.gla.ac.uk/Slavonic/Czech_Hist4.html). Arts.gla.ac.uk. . Retrieved 25 April 2010.

[22] *" Plague epidemics in Czech countries* (http://books.google.com/books?id=3u3rNCWtv0MC&pg=PA49&dq&hl=en#v=onepage&q=& f=false)". E. Strouhal. p.49.

[23] Oskar Krejčí, Martin C. Styan, Ústav politických vied SAV. (2005). *Geopolitics of the Central European region: the view from Prague and Bratislava* (http://books.google.com/books?id=38ciAe4J4VMC&pg=&dq&hl=en#v=onepage&q=&f=false). p.293. ISBN 80-224-0852-2

[24] "RP's History Online – Habsburgs" (http://archiv.radio.cz/history_96/history07.html). Archiv.radio.cz. . Retrieved 25 April 2010.

[25] " *History of the Mongols from the 9th to the 19th Century. Part 2. The So-Called Tartars of Russia and Central Asia. Division 1* (http:// books.google.com/books?id=j08L5xLOQKwC&pg=PA557&dq&hl=en#v=onepage&q=&f=false)". Henry Hoyle Howorth. p.557. ISBN 1-4021-7772-0

[26] " *The new Cambridge modern history: The ascendancy of France, 1648–88* (http://books.google.com/books?id=FzQ9AAAAIAAJ& pg=PA494&dq&hl=en#v=onepage&q=&f=false)". Francis Ludwig Carsten (1979). p.494. ISBN 0-521-04544-4

[27] "*The Cambridge economic history of Europe: The economic organization of early modern Europe*". E. E. Rich, C. H. Wilson, M. M. Postan (1977). p.614. ISBN 0-521-08710-4

[28] Radio Praha – zprávy (http://www.radio.cz/cz/zpravy/105664) (30 June 2008) (Czech)

[29] "Tab. 3 Národnost československých státních příslušníků podle žup a zemí k 15.2.1921" (http://web.archive.org/web/20070605105429/ http://www.czso.cz/sldb/sldb.nsf/i/8BE4678613181F2AC1256E66004C77DD/$File/tab3_21.pdf) (in Czech) (PDF). Czech Statistical Office. Archived from the original (http://www.czso.cz/sldb/sldb.nsf/i/8BE4678613181F2AC1256E66004C77DD/$File/tab3_21.pdf) on 5 June 2007. . Retrieved 2 June 2007.

[30] Gerhard L. Weinberg, *The Foreign Policy of Hitler's Germany: Starting World War II, 1937–1939* (Chicago, 1980), pp. 470–481.

[31] Stephen A. Garrett (1996). " *Conscience and power: an examination of dirty hands and political leadership* (http://books.google.com/ books?id=lF9NBUZlh0AC&pg=PA60&dq&hl=en#v=onepage&q=&f=false)". Palgrave Macmillan. p.60. ISBN 0-312-15908-0

[32] " *A Companion to Russian History* (http://books.google.com/books?id=JyN0hlKcfTcC&pg=PA409&dq&hl=en#v=onepage&q=& f=false)". Abbott Gleason (2009). Wiley-Blackwell. p.409. ISBN 1-4051-3560-3

[33] F. Čapka: Dějiny zemí Koruny české v datech (http://www.libri.cz/databaze/dejiny/text/t98.html). XII. Od lidově demokratického po socialistické Československo – pokračování. Libri.cz (Czech)

[34] Czech schools revisit communism (http://news.bbc.co.uk/2/hi/europe/4388764.stm). *BBC News*. 1 November 2005.

[35] "Klaus signs Czech direct presidential election implementing law" (http://www.ceskenoviny.cz/tema/zpravy/ klaus-signs-czech-direct-presidential-election-implementing-law/823441). Czech Press Agency. 2012-08-01. . Retrieved 2012-11-07.

[36] "Czechs with few mates" (http://www.economist.com/displayStory.cfm?story_id=9725352). *The Economist*. 30 August 2007. . Retrieved 25 April 2010.

[37] The death of the districts (http://www.radio.cz/en/article/36046), Radio Prague 3 January 2003.

[38] R. Tolasz, *Climate Atlas of Czechia*, Czech Hydrometeorological Institute, Prague, 2007. ISBN 8024416263, graphs 1.5 and 1.6

[39] "Czech absolute record temperature registered near Prague" (http://www.ceskenoviny.cz/news/zpravy/ czech-absolute-record-temperature-registered-near-prague/830626). *České noviny*. ČTK. . Retrieved 20 August 2012.

[40] R. Tolasz, *Climate Atlas of Czechia*, Czech Hydrometeorological Institute, Prague, 2007. ISBN 8024416263, graph 2.9.

[41] Getting to know Czech Republic (http://www.czech.cz/en/czech-republic/getting-to-know-czech-republic/), from Czech.cz (http:// www.czech.cz/), the official site of the Czech Republic

[42] "World Bank 2007" (http://web.worldbank.org/WBSITE/EXTERNAL/DATASTATISTICS/ 0,,contentMDK:20421402~pagePK:64133150~piPK:64133175~theSitePK:239419,00.html#High_income). Web.worldbank.org. . Retrieved 25 April 2010.

[43] "GDP per capita in PPS" (http://epp.eurostat.ec.europa.eu/cache/ITY_PUBLIC/2-25062009-BP/EN/2-25062009-BP-EN.PDF). Eurostat. . Retrieved 25 June 2009.

[44] Stastny, Daniel (2010). "Czech Economists on Economic Policy: A Survey" (http://econjwatch.org/issues/ volume-7-issue-3-september-2010). *Econ Journal Watch* **7** (3): 275–287. .

[45] "Czech Republic to join Schengen" (http://web.archive.org/web/20080225173344/http://www.praguepost.com/articles/2006/12/13/ czech-republic-to-join-schengen.php). The Prague Post. 13 December 2006. Archived from the original (http://www.praguepost.com/ articles/2006/12/13/czech-republic-to-join-schengen.php) on 25 February 2008. . Retrieved 8 October 2007.

[46] "Finance Ministry backtracks on joining the Euro by 2012" (http://www.radio.cz/en/news/94849). Radio Praha. . Retrieved 22 December 2008.

[47] "Czech government adopts euro adoption plan" (http://web.archive.org/web/20070930165055/http://www.eubusiness.com/Euro/ czech-euro.83/). EUbusiness. 11 April 2007. Archived from the original (http://www.eubusiness.com/Euro/czech-euro.83/) on 30 September 2007. . Retrieved 1 June 2007.

[48] "For Czechs, euro adoption still a long way off" (http://www.radio.cz/en/section/curraffrs/ for-czechs-euro-adoption-still-a-long-way-off). Radio.cz. 17 June 2012. . Retrieved 19 August 2012.

[49] "Range of rank on the PISA 2006 science scale" (http://www.oecd.org/dataoecd/42/8/39700724.pdf) (PDF). *OECD.org*. . Retrieved 25 April 2010.

[50] "Transport infrastructure at regional level – Statistics explained" (http://epp.eurostat.ec.europa.eu/statistics_explained/index.php/ Transport_infrastructure_at_regional_level#Railways). Epp.eurostat.ec.europa.eu. . Retrieved 25 April 2010.

[51] "Railway Network in the Czech Republic" (http://provoz.szdc.cz/portal/Show.aspx?oid=185802). *SZDC.cz*. . Retrieved 9 November 2010.

[52] (Czech) Roads and Motorways in the Czech Republic (http://www.rsd.cz/rsd/rsd.nsf/0/80345976071FCBACC12575CF004E133E/
$file/RSD2009en.pdf). RSD.cz (2009).

[53] (http://www.rsd.cz/sdb_intranet/sdb/download/prehledy_2011_1_cr.pdf)

[54] Lee Taylor (2 May 2012). "'State of the Internet' report reveals the fastest web speeds around the world" (http://www.news.com.au/
technology/state-of-the-internet-reports-reveals-internet-speeds-around-the-world/story-e6frfro0-1226344876391). *news.com.au*. . Retrieved
2 May 2012.

[55] 2007 WiFi survey EN (http://www.ey.com/global/content.nsf/Czech_Republic_E/2007_WiFi_survey_EN)

[56] "Openspectrum.info – Czech Republic" (http://liveweb.archive.org/http://www.volny.cz/horvitz/os-info/czech.html). Volweb.cz. .
Retrieved 25 April 2010.

[57] "Wi-Fi: Poskytovatelé bezdrátového připojení" (http://translate.google.com/translate?u=http://www.internetprovsechny.cz/
wifi-poskytovatele.php&hl=cs&ie=UTF8&sl=cs&tl=en). internetprovsechny.cz. . Retrieved 17 March 2008.

[58] "Bezdrátové připojení k internetu" (http://translate.google.com/translate?u=http://www.bezdratovepripojeni.cz&hl=cs&ie=UTF8&
sl=cs&tl=en). bezdratovepripojeni.cz. . Retrieved 18 May 2008.

[59] Ingenious inventions (http://web.archive.org/web/20090324033030/http://www.czech.cz/en/economy-business-science/science/
ingenious-inventions?i=1). Czech.cz. Retrieved 3 March 2009.

[60] The History of Contact Lenses (http://www.eyetopics.com/articles/18/1/The-History-of-Contact-Lenses.html). Retrieved 3 March
2009.

[61] "Velikáni české vědy" (http://www.velikani-ceske-vedy.estranky.cz/clanky/biografie/stanislav-brebera). . Retrieved 1 November 2010.

[62] "Faces of the Presidency" (http://www.eu2009.cz/en/czech-presidency/presidency-faces/faces-of-the-presidency-3642/). *eu2009.cz*.
EU2009.cz. . Retrieved 8 January 2009.

[63] "Tomio Okamura: Praha je 6. nejnavštěvovanější město Evropy – Televize plná Prahy" (http://www.metropol.cz/zpravy/z-prahy/
tomio-okamura-praha-je-6-nejnavstevovanejsi-mesto-evropy/). Metropol.cz. . Retrieved 27 May 2012.

[64] "Promotion Strategy of the Czech Republic in 2004–2010" (http://web.archive.org/web/20070328141615/http://www.czechtourism.
com/index.php?show=001006&lang=3). Czech Tourism. Archived from the original (http://www.czechtourism.com/index.
php?show=001006&lang=3) on 28 March 2007. . Retrieved 19 December 2006.

[65] "Prague sees significant dip in tourist numbers" (http://www.radio.cz/en/article/106877). Radio.cz. 21 April 2010. . Retrieved 25 April
2010.

[66] Prague mayor goes undercover to expose the great taxi rip-off (http://www.independent.co.uk/news/world/europe/
prague-mayor-goes-undercover-to-expose-the-great-taxi-ripoff-486682.html), 15 January 2005

[67] Tips on Staying Safe in Prague (http://www.myczechrepublic.com/prague/safety.html), myczechrepublic.com

[68] Czech Republic – Country Specific Information (http://travel.state.gov/travel/cis_pa_tw/cis/cis_1099.html#crime), U.S. Department of
State

[69] "Aquapalace Praha bude největším aquaparkem ve střední Evropě" (http://www.konstrukce.cz/clanek/
aquapalace-praha-bude-nejvetsim-aquaparkem-ve-stredni-evrope/). Konstrukce.cz. . Retrieved 27 May 2012.

[70] "2011 census" (http://notes2.czso.cz/cz/sldb2011/cd_sldb2011_11_12/kraje.html) (in Czech). Czech Statistical Office. . Retrieved 8
May 2012.

[71] "Statistická data ČSÚ (ČSÚ statistical data)" (http://www.statnisprava.cz/) (in Czech). Czech Statistical Office. . Retrieved 8 May 2012.

[72] Nařízení vlády č. 212/1997, kterým se vyhlašuje závazná část územního plánu velkého územního celku Olomoucké aglomerace (http://
aplikace.mvcr.cz/sbirka-zakonu/ViewFile.aspx?type=c&id=3066)

[73] Územní plán velkého územního celku ČESKOBUDĚJOVICKÉ SÍDELNÍ AGLOMERACE, (http://up.kraj-jihocesky.cz/files/
up_vuc_cbsra[1].pdf)

[74] První předběžné výsledky Sčítání lidu, domů a bytů 2011: Obyvatelstvo podle národnosti podle krajů (http://notes2.czso.cz/cz/sldb2011/
cd_sldb2011_11_12/index_html_files/PVCR062.pdf). (PDF) . Retrieved on 12 August 2012.

[75] "The History and Origin of the Roma" (http://romove.radio.cz/en/article/18158). Romove.radio.cz. . Retrieved 25 April 2010.

[76] Green, Peter S. (5 August 2001). "British Immigration Aides Accused of Bias by Gypsies" (http://query.nytimes.com/gst/fullpage.
html?res=9401E6DA103CF936A3575BC0A9679C8B63&sec=&spon=&pagewanted=all). *New York Times*. . Retrieved 25 April 2010.

[77] Foreigners by type of residence, sex and citizenship (http://www.czso.cz/csu/cizinci.nsf/engt/8200578577/$File/c01t01.pdf), *Czech
Statistical Office*, 31 October 2009

[78] "The Holocaust in Bohemia and Moravia" (http://www.ushmm.org/wlc/article.php?lang=en&ModuleId=10007323). Ushmm.org. .
Retrieved 25 April 2010.

[79] The Virtual Jewish Library (http://www.jewishvirtuallibrary.org/jsource/Judaism/jewpop.html) – Jewish population of Czech republic,
2005

[80] " PM Fischer visits Israel (http://www.radio.cz/en/article/118537)". Radio Prague. 22 July 2009.

[81] CIA World Factbook (https://www.cia.gov/library/publications/the-world-factbook/rankorder/2127rank.html?countryName=Czech
Republic&countryCode=ez®ionCode=eur&rank=215#ez): Total Fertility Rate ranks

[82] " Press: Number of foreigners in ČR up ten times since 1989 (http://groups.yahoo.com/group/scotch-irish/message/50884)". Prague
Monitor. 11 November 2009.

[83] O'Connor, Coilin (29 May 2007). "Is the Czech Republic's Vietnamese community finally starting to feel at home?" (http://www.radio.cz/
en/article/91826). *Czech Radio*. . Retrieved 1 February 2008.

[84] Crisis Strands Vietnamese Workers in a Czech Limbo (http://www.nytimes.com/2009/06/05/world/europe/05iht-viet.
      html?pagewanted=1&_r=1). *The New York Times*. 5 June 2009.
[85] " Foreigners working in the Czech Republic (http://web.archive.org/web/20090603021134/http://www.czech.cz/en/current-affairs/
      work-and-study/foreigners-working-in-the-czech-republic)". Ministry of Foreign Affairs. July 2006.
[86] Czechs and Bohemians (http://www.encyclopedia.chicagohistory.org/pages/153.html). Encyclopedia of Chicago.
[87] Czech and Slovak roots in Vienna (http://www.wieninternational.at/en/node/3586). Wieninternational.at
[88] "U.S. Census" (http://factfinder.census.gov/servlet/ADPTable?_bm=y&-geo_id=01000US&-ds_name=ACS_2006_EST_G00_&
      -_lang=en&-_caller=geoselect&-format=). U.S. Census Bureau. . Retrieved 13 April 2008.
[89] "Population by religious belief and by municipality size groups" (http://www.czso.cz/sldb2011/eng/redakce.nsf/i/
      tab_7_1_population_by_religious_belief_and_by_municipality_size_groups/$File/PVCR071_ENG.pdf). Czech Statistical Office. .
      Retrieved 23 April 2012.
[90] Richard Felix Staar, *Communist regimes in Eastern Europe*, Issue 269, p. 90
[91] "Population by denomination and sex: as measured by 1921, 1930, 1950, 1991 and 2001 censuses" (http://www.czso.cz/csu/
      2008edicniplan.nsf/engt/24003E05ED/$File/4032080119.pdf) (in Czech and English). Czech Statistical Office. . Retrieved 9 March 2010.
[92] "Eurobarometer on Social Values, Science and technology 2005 – page 11" (http://ec.europa.eu/public_opinion/archives/ebs/
      ebs_225_report_en.pdf) (PDF). . Retrieved 5 May 2007.
[93] "Social values, Science and Technology" (http://ec.europa.eu/public_opinion/archives/ebs/ebs_225_report_en.pdf) (PDF).
      Eurobarometer. June 2005. . Retrieved 19 December 2006.
[94] "KFTV" (http://www.kftv.com/country/Czech_Republic/guide/general). Wilmington Publishing and Information Ltd. . Retrieved 26
      October 2012.
[95] "Czech Film Commission - Karlovy Vary" (http://www.filmcommission.cz/news/detail/id/85). Czech Film Commission. . Retrieved 26
      October 2012.
[96] (Czech) Kyša, Leoš (January 28, 2011). "Počet legálně držených zbraní v Česku stoupá. Už jich je přes 700 tisíc" (http://domaci.ihned.
      cz/c1-49617340-pocet-legalne-drzenych-zbrani-v-cesku-stoupa-uz-jich-je-pres-700-tisic). ihned.cz. . Retrieved January 28, 2011.
[97] Jagr Scores 1,600th Point in Return (http://www.nytimes.com/2011/10/07/sports/hockey/nhl-hockey-roundup.html?_r=1). Associated
      Press via The New York Times (7 October 2011).

Further reading

- Hochman, Jiří. *Historical dictionary of the Czech State* (1998)

External links

- Czech Republic (http://www.czech.cz/en/).
- (http://www.archeolog.cz/encyklopedie/keltove-kelt-keltske/)
- (http://www1.ceses.cuni.cz/benacek/hist kniha2.pdf)

Government

- Governmental website (http://www.vlada.cz/en/).
- Presidential website (http://www.hrad.cz/en/).
- Portal of the Public Administration (http://portal.gov.cz/portal/obcan/).
- Senate (http://www.senat.cz/index-eng.php).
- Chief of State and Cabinet Members (https://www.cia.gov/library/publications/world-leaders-1/
  world-leaders-c/czech-republic.html).

General information

- Czech Republic (https://www.cia.gov/library/publications/the-world-factbook/geos/ez.html) entry at *The
  World Factbook*
- Czech Republic (http://www.state.gov/p/eur/ci/ez/) information from the United States Department of State.
- Portals to the World (http://www.loc.gov/rr/international/european/czechr/cz.html) from the United States
  Library of Congress.
- Czech Republic (http://ucblibraries.colorado.edu/govpubs/for/czechrepublic.htm) at *UCB Libraries
  GovPubs*.
- Czech Republic (http://www.dmoz.org/Regional/Europe/Czech_Republic/) at the Open Directory Project
- Czech Republic profile (http://www.bbc.co.uk/news/world-europe-17220018) from the BBC News

- Wikimedia Atlas of the Czech Republic
- Geographic data related to Czech Republic (http://www.openstreetmap.org/browse/relation/51684) at OpenStreetMap
- Key Development Forecasts for the Czech Republic (http://www.ifs.du.edu/ifs/frm_CountryProfile. aspx?Country=CZ) from International Futures

News

- The Prague Post (http://www.praguepost.com/).
- CzechNews (http://aktualne.centrum.cz/czechnews/).
- Czech News Agency News (http://www.ceskenoviny.cz/news/).
- Prague Daily Monitor (http://praguemonitor.com/).
- Radio Prague (http://www.radio.cz/en/).

Statistics

- Czech Statistical Office (http://www.czso.cz/eng/redakce.nsf/i/home).

Travel

- Czech Tourism (http://www.czechtourism.com/Homepage.aspx?lang=en-GB&selectedculture=en-US) Official travel site of the Czech Republic.

Kunčice_(Hradec_Králové_District)

Kunčice	
Village	
Country	Czech Republic
Region	Hradec Králové
District	Hradec Králové
Commune	Hradec Králové
Municipality	Nechanice
Elevation	262 m (860 ft)
Coordinates	50°13′3″N 15°38′11″E
Area	5.91 km^2 (2.28 sq mi)
Population	257 (*2006-10-02*)
Density	43 / km^2 (111 / sq mi)
First mentioned	1382
Mayor	Květoslav Kvasnička
Timezone	CET (UTC+1)
- summer (DST)	CEST (UTC+2)
Postal code	503 15

Location in the Czech Republic

Statistics: statnisprava.cz [1]

Website: www.kuncice.info [2]

Kunčice is a village in the Czech Republic.

External links

- Municipal website [2]

References

[1] http://www.statnisprava.cz/ebe/ciselniky.nsf/i/570214
[2] http://www.kuncice.info/

Article Sources and Contributors

Kulíľov *Source*: http://en.wikipedia.org/w/index.php?title=Kul%C3%AD%C5%99ov *Contributors*:

Village *Source*: http://en.wikipedia.org/w/index.php?title=Village *Contributors*: 1223Sallybride, 21655, A.Hassen, AKeen, Aarohan Mettu, Abductive, Acsian88, AdamDeanHall, Adamhauner, Adityamadhav83, Airjatt, Aitias, Akrabbim, Al Ameer son, Alaexis, Ale jrb, AlefZet, Aleksandr Grigoryev, Alensha, Altenmann, Amaury, Amberrock, Aminullah, Andrew M. Vachin, Andrewpmk, Andycjp, Animum, Apalamarchuk, Apokrif, Arpingstone, Arthena, AussieLegend, Azreey, Bejinhan, Beland, Belirac, Berig, Big Bird, Bill37212, Biruitorul, Bjh21, Bkonrad, Bldonne, Bobrayner, BocoROTH, Bogdan, Bogdangiusca, Bomac, Bossmanham13, Bossmanham98, Branka France, BreakingFree40, Brian the Editor, Brian0918, Buistr, Butko, C.Fred, Caiaffa, CalJW, CalumMcA, Canley, Capricorn42, Cassowary, Catgut, Catslisten, Cecropia, Cenarium, Ceriy, Cgoodwin, Chasnor15, Chentchen, Chirag, Chris j wood, Christopher.e.dunn, Chzz, Ciudad jardin, Cntras, Colonies Chris, CommonsDelinker, Conleyb15, Conorbrady.ie, Corregere, Cosmic Latte, CrimsonBlue, Cwolfsheep, D6, DITWIN GRIM, Da-drumer, DailyWikiHelp, Daniel Case, Daniel J. Leivick, Daniel Quinlan, David Biddulph, David Kernow, Dbg1233, Ddstretch, DePiep, Deathawk, Deeptrivia, Den fjättrade ankan, Deqon, Deror avi, Desiphral, Dezaree, Dieter Simon, Dinesh.jangra1, Discospinster, Djr13, Doctor Whom, Donarreiskoffer, Doseiai2, Dottod, Dpkipling, Durgesh.Rai, Dysepsion, E ponti, E0steven, Egard89, Ehrenkater, Elassint, Eleassar, Ellywa, Elockid, Elroygoh02, Emyr93, Epipelagic, Erianna, Eu.stefan, EugeneZelenko, EurekaLott, Extra999, Ezhiki, Fang 23, Farosdaughter, Fat pig73, Fayenatic london, Flewis, Francis Tyers, Francs2000, Fred Bauder, Frosted14, Fujurcitook, Funandtrvl, FunkyFly, Futurebird, Gardar Rurak, Garhowell, Gegnome, Gfoley4, Gingermint, Ginkgo100, Giro720, Glenn, Globalphilosophy, GoOdCoNtEnT, Gogo Dodo, Gorai, Grafikm fr, GrahamSmithe, Gtswiki, Guanaco, Guss2, Gustavo Szwedowski de Korwin, Gwen-chan, Gz33, Hahnchen, Halsteadk, Harisu1988, Hayabusa future, Headtoadman, Hebrides, Helennwarwick, HenryLi, HighSpeed-X, Hmains, Hohum, Hroðulf, Hu12, Hydrogen Iodide, IGeMiNix, IShadowed, IW.HG, Ianisbored, Indiaforever, Ineffable3000, Inwind, Isam, IstvanWolf, IvanKlinko, J.delanoy, JHunterJ, JPSheridan, Jahangard, James086, Janetleopold, Jared Preston, Jcuk, JdeJ, Jeremy Bolwell, Jerrch, Jim1138, Jno, Jongleur100, Julesd, Julia W, JuneGloom07, KR3ColinB, Kajasudhakarababu, Kamiekam, Keelanrush, Kelisi, Kelovy, Khalid Mahmood, Khanvez, Khazar2, Kikos, Kimberry352, Kironbd07, KirrVlad, Kjramesh, Kman543210, Kmsasidhar, Komalrajiana, Ktr101, Kurruption, Kwamikagami, Larucio, Laurinavicius, Ldonna, Leonig Mig, Levineps, Lew121, LiDaobing, Lightmouse, LilHelpa, Linaduliban, Little muddy funkster, Littlealien182, Llywrch, Lonewolf BC, Lord Geznikor, Lozleader, Luk, Lulu1127, Lumos3, Lynda Finn, MacTire02, Maglame, Maitch, Majorly, Malick78, Mallakottai, Mani1, Manly 2008, Marek69, Mariano12 1989, Mariwan zanko, Mark91, Markmark12, Materialscientist, Mayooranathan, Mdzafri, Meldor, Menacheryarun, Mfo7787, MickMacNee, Mike Rosoft, MisfitToys, MisterVodka, Modulatum, Moebiusuibeom-en, Moonriddengirl, Mrmuk, Ms. Hayden and Ms. S, Mtaylor848, Mvjs, Mwalcoff, Mxn, NE2, Naturenet, Naveen Sankar, NerdyScienceDude, Nethergreen55, NewEnglandYankee, Nickshanks, Nistra, Noclevername, Northdeigo, Nricardo, Nyttend, Olive, Old Moonraker, Orangemike, OscarKosy, P.Marlow, Pankajkumar.9009, Parkwells, Patrick, PaulBannister, Pawyilee, Penrithguy, Peter James, PeterCScott, Pgan002, Phagopsych, Pharaoh of the Wizards, PhilKnight, Pinethicket, PleaseStand, Pokrajac, Ppeace, Pug1776, Puneet 348, Quaysys, R'n'B, RDF, Raeky, Randhirreddy, RandomAct, Rcingham, Realteacher, Reeses14, Reinyday, Rich Farmbrough, Rickyrab, Risker, Rjm at sleepers, Robdav69, Rocket71048576, Rwxrwxrwx, Saccerzd, Sadrettin, Saga City, Same7same7, SamuraiClinton, Samyo, Sansonic, Sardanaphalus, Schinhofen, Schzmo, ScottSteiner, Seb az86556, Seivadnadroj, Sem10efg, Sephiroth BCR, Sfu, Shakingjoker117, Shamsurabiu, Shantavira, Sheddybey, Shell Kinney, Shergo korali, Shoeofdeath, SilkTork, Sjakkalle, Sk jmp, Sketchmoose, Sljaxon, Sole Soul, Speciate, StephenDawson, Steven Zhang, Stevo D, StewartNetAddict, Stifle, Student BSMU, Subhrat, Suffusion of Yellow, Suprgye, Surya Prakash.S.A., Szambo, THATSBETTER, Taamu, Tabletop, TaintedMustard, Tangerinehistry, Tbhotch, Teinesavaii, TejasDiscipulus2, TenIslands, Tenth Plague, TerryAga, The Thing That Should Not Be, The Transhumanist, TheArguer, Theranos, This, that and the other, Thomasrflagel, Tide rolls, Tjanke, Tobby72, Tobias Conradi, Tpbradbury, Triona, Tsuchiya Hikaru, TurkChan, Twanner921, Twin Bird, Tytyty456, Ukabia, UnicornTapestry, Untifler, Utcursch, V2k, VMS Mosaic, Vadim Kiev, Vegaswikian, Vhdwikiuser, Vikramdidawat, Vivek Rai, Vmenkov, WOSlinker, Walex03, Warofdreams, Wayland, Wayne Slam, Wendy1991, West Brom 4ever, Whytecypress, Widr, Wiki Wikardo, Wikijens, William Allen Simpson, William Avery, Winhunter, Wknight94, WojPob, Woohookitty, WorldWide Update, Wrelwser43, Xiaoyu of Yuxi, Xyz or die, Yakudza, Yonkie, Yupik, Zacharie Grossen, Zloyvolsheb, Zoeyzazu, Zollerriia, Zotel, Александър, ქართული მ, , , 472 anonymous edits

Obec *Source*: http://en.wikipedia.org/w/index.php?title=Obec *Contributors*: Chanheigeorge, Dzyadyk, Europravnik, Haddiscoe, JAn Dudík, Jetam2, Juro, MarkBA, Miaow Miaow, Rich Farmbrough, Tobias Conradi, ŠJů, 10 anonymous edits

Human_settlement *Source*: http://en.wikipedia.org/w/index.php?title=Human_settlement *Contributors*: Andrewaskew, Andycjp, Ankuthakur848, Anthony Appleyard, Arunsingh16, Bjankuloski06en, Bjenks, Bkonrad, Bogdan Nagachop, CWii, CambridgeBayWeather, Canterbury Tail, Cometstyles, CrimsonBlue, D.M. from Ukraine, Deagle AP, Donald Albury, Drenaline, Edderso, Elekhh, Epbr123, Euchiasmus, Ewlyahoocom, Feministo, Footballexpert, Foxj, Fusion7, Galoubet, Ganpop, Glane23, Glenn, Grafen, Hmains, Hwy43, Inwind, JSpung, JamesBWatson, Jauhienij, Jellyman, Joshua Issac, Joy, Jurema Oliveira, KR3ColinB, Kotniski, Levineps, Lobsterthermidor, Look2See1, Maudemiller, Mayooranathan, Mazadillon, Mdd, Meco, Mike Christie, Mike hayes, Mindstalk, My76Strat, NERIUM, Newyorkbrad, Nick Number, Patchy1, Rjm at sleepers, Rjwilmsi, RockfangSemi, Saddhiyama, Saga City, Smk1895x, SomeOtherInsect, Spencer, TenIslands, Tertulius, The Aviv, TheAllSeeingEye, TigerPaw2154, Tokek, Vanished user ewfisn2348tui2f8n2fio2utjfeoi210r39jf, Varjacic Vladimir, Vini sharma, Wavelength, Whytecypress, Wikiwatcher19, Wlodzimierz, Woohookitty, Xofc, Yerpo, Yuma en, Zzyxzaa26, Мурад 97, يارابم 1971, 90 anonymous edits

Nymburk_District *Source*: http://en.wikipedia.org/w/index.php?title=Nymburk_District *Contributors*: Dr. Blofeld, Ser Amantio di Nicolao, Shadster

Central_Bohemian_Region *Source*: http://en.wikipedia.org/w/index.php?title=Central_Bohemian_Region *Contributors*: Amikake3, BD2412, Bejnar, Bggoldie, Caroig, Chanheigeorge, Cloudz679, Darkwind, Darwinek, David Kernow, Dbachmann, Delasoto, Dr. Blofeld, Elapsed, Epolk, Grutness, Gryffindor, Hareco, Jfblanc, John Quincy Adding Machine, Knepflerle, Krokodyl, Lectonar, Lmach, Lokolskaia, Marek69, Matt Borak, Matthead, Maximus Rex, Miaow Miaow, Mirekk.cz, Neitherday, Okino, Olessi, Olivier, Orioane, PZFUN, Palica, Pavel Vozenilek, Piccolo Modificatore Laborioso, Ppntori, PrcekA, Qertis, Rarelibra, Riso, Sapfan, Sheynhertz-Unbayg, Starky, Template namespace initialisation script, Tokiohotelover, Urmas, Vlad Kacer, Voyevoda, Whhalbert, Zeman, ქართული მ, 24 anonymous edits

Bříství *Source*: http://en.wikipedia.org/w/index.php?title=B%C5%99%C3%ADstv%C3%AD *Contributors*: D6, Darwinek, Dr. Blofeld, Karelj, Rich Farmbrough, Ser Amantio di Nicolao

Bohemia *Source*: http://en.wikipedia.org/w/index.php?title=Bohemia *Contributors*: -Ilhador-, 12.47, 52 Pickup, Adamkraft, Aecis, Aelffin, Aetheling1125, Agentv, Aitias, AjaxSmack, Alan Flynn, Albedo, Alex.muller, Alexander Domanda, Altenmann, Alxeedo, Angela, Angusmclellan, Animum, Antakyan, Anthony Appleyard, Aramgutang, Argonauth, Arminius09, Arthena, Attilios, Ausir, Badcorp, Barticus88, Beardo, Berrys70, Binabik80, Biruitorul, Bjankuloski06en, Bletch, Bluedenim, Bogdan, Bovineone, Bradjohns10, CALR, CSvBibra, Caknuck, CalJW, Caltas, Cathack, Cautious, Cepek, Charles Matthews, CharlotteWebb, ChrisCork, Christopher Frey, Clarkcorp, Commander Keane, CommonsDelinker, Conversion script, Cremepuff222, CrniBombarder!!!, Daarznieks, Davewho2, David Kernow, David Parker, Dbachmann, Delasoto, Den fjättrade ankan, DesmondW, Discospinster, DocWatson42, Domino theory, Dominus Vobisdu, Donatus, Dpol, Drmissio, Dryazan, Dsal, Dysprosia, Eclecticology, Ed Poor, Elassint, EmilJ, Empetl, Enfantsduparadis, Erendwyn, Erianna, EugeneZelenko, Ev, Evanmaro, Evilive666, Excirial, Fabartus, Favonian, Feix, Fig wright, Flibirigit, Flibjib8, Fransvannes, Fred Bauder, FredR, Frokor, Funfunfun23, Fwb22, GSTQ, Gabbe, Gaius Cornelius, Gary King, Gazilion, Gem-fanat, Gigemag76, Glennimoss, Gresszilla, Gryffindor, Gurch, H.J., HamburgerRadio, Hanmleteer, Herk1955, Hesperian, Hetar, Hibernian, Hilarius boogerdam, Hmains, Honzula, Hutcher, IEEE, Ianweller, Iaroslavvs, Ikalmar, Imaginatorium, Immunize, Inb4thelock, Iohannes Animosus, Iricigor, Itai, Ivan Štambuk, J. Spencer, J04n, JCSantos, JHK, JMvanDijk, JSoules, JamieS93, Jan.Kamenicek, Jararmando, Jaromír-A, Jeff G., Jeffq, Jennavecia, JesseGarrett, JialiangGao, Jim10701, Jim1138, Jirka6, Jklamo, John of Reading, Jojit fb, Joostik, Joseph Solis in Australia, Joshua Scott, Joy, Jsc83, Junckerg, Juro, Jwoodger, Karasek, Katieh5584, Kaz, Keilana, Kick3932os, Kielsky, King Harald II of Bohemia, Kingpin13, KirkEN, Kozuch, Kyle1278, Lars T., Lhartranft, Lightmouse, Lilac Soul, Macaddct1984, Maccakennet, MacedonianBoy, Makemi, Malhamboy, Marek69, Mark Renier, Markussep, Marzalpac, Mat-C, Materialscientist, Matt Heard, Matthead, Maus-78, Mav, MayerG, Menchi, Mentifisto, Merenta, Meursault2004, Michael Zimmermann, Mikeo, Mikewebkist, Miller17CU94, Mintleaf, Mogism, Molobo, Mondo one, MrPMonday, Naive cynic, Neilbeach, Nick, NickBush24, Nico, NicoNet, Niteowlneils, Nk, Oashi, Ohconfucius, Ojis, Olessi, Olivier, Omnipaedista, Oneliner, Ossipewsk, OwenBlacker, PANONIAN, Pajast, Panscient, Paulmlieberman, Pavel Vozenilek, Peregrine981, Peter Isotalo, Pgdudda, Pigman, Piotrus, Porcher, PowerCS, Primus128, Pseudo-Richard, Punkturnip, Qertis, RG2, RJaguar3, Raeky, RandomCritic, Reo On, Rich Farmbrough, Road Wizard, Robertgreer, Rshu, SMP, SURIV, SWAdair, Sam Hocevar, Sandius, Saxifrage, SchfiftyThree, SebastianBreier, Seelo, Shanedidona, Shenme, SigmaEpsilon, Sindinero, Skenmintz, Smith2006, Snowdog, Soelke, Space Cadet, Squash Racket, StanZegel, Steeev, Stijn Calle, Suisui, Susfele, Svetovid, TFMcQ, Tdjewell, TexasAndroid, TheWhiteRussian, Thecrystalcicero, Thomas Graves, Thumperward, Tikiwont, Tom harrison, Tomfutru, Tommy2010, Tonyfaull, Trilobite, Trusilver, Tulkolahten, Ultor Solis, Ultratomio, UtherSRG, Varlaam, Vitloh, WAS 4.250, Wachowich, Wee Jimmy, Wiglaf, Wiki alf, Wikimol, William Pembroke, Wizzard, Woohookitty, Xmort, Yamamoto Ichiro, Yidisheryid, Yopie, Yuyu, Zingus, Zsinj, 477 anonymous edits

Czech_Republic *Source*: http://en.wikipedia.org/w/index.php?title=Czech_Republic *Contributors*: .hjkyhuikrtdh, -- April, -Majestic-, -jkb-, 100110100, 1exec1, 23prootie, 2A02:2F01:1059:F004:0:0:BC19:A31A, 2help, 334a, 4cecz4, 5 albert square, A Werewolf, A bit iffy, Abcbc, Acroterion, Adam J. Sporka, Adam mugliston, Adam78, Adammathias, Addshore, Aetheling1125, Aherunar, Ahoerstemeier, Aika-sáni, Aitias, Aivazovsky, AjitPD, Akai Goth, Akanemoto, Aktron, Alaiche, Alansohn, Alarob, Alawadhi3000, AlefZet, AlexiusHoratius, Alfie66, Alfvaen, Alkari, Alois Musil, Alphachimp, Alphasinus, Altenmann, Amakuru, Amaury, Ananaso, Andem, AndreasJS, Andrew Gwilliam, Andrewlp1991, Andrwsc, Andy Marchbanks, Andycjp, Angelo De La Paz, Anger22, Animum, AnnaFrance, Antandrus, Anthonyd3ca, Antiuser, Appleseed, Aquilina, Aquilla, ArchonMeld, ArglebargleIV, Argyriou, Arnon Chaffin, Arouze, Arsenic99, Art LaPella, Arwel Parry, Ashmoo, Ashxhole, Asidemes, Assdffghjkkkkl, AstroNomer, Atitarev, Atoric, Atwardow, Augustes, Auror, Avala, Avenged Eightfold, AxG, Ayla, AzaToth, BRG, Baa, BalkanFever, Banaticus, Barek, Baronnet, Barryob, Bart133, Bazonka, Bbarkley2, Bbisdo, Bcnviajero, Becherka, Beetstra, Bemoeial, Bermicourt, BernardaAlba, Beyond silence, Bg007, Bibliomaniac15, Bidabadi, Big Adamsky, Biker Biker, Bill3000, Billinghurst, Birion, Biruitorul, Bjarki S, Bjkeys321, Bkell, Black&White, BlaiseFEgan, Blakeysfc, Blanicky, Blood Code ABACABB, Blue-Haired Lawyer, BlueMars, Bluemask, Bluewind, Bluezy, Bob Burkhardt, Bobblehead, Bobo192, Bohuslavroztocil, Boing! said Zebedee, Bokpasa, Bombonek, Bombs Bombs Away!, Bonadea, Boothy443, Bože pravde, BradBeattie, Brandmeister, BrendelSignature, Bribri420, Brion VIBBER, BritishWatcher, Brumski, Brynp, Bsadowski1, Btball, Buaidh, C.Fred, C9900, CART fan, Cacuija, Caesura, Caknuck, Calaschysm, CalicoCatLover, Caltas, Camembert, Camw, Can't sleep, clown will eat me, Canthusus, Cantrelllee, Cantus, Canuckian89, Caponer, Caroig, Cbradshaw, Cembo123, Cepek, ChKa, Chase me ladies, I'm the Cavalry, Chasingsol, Chavash, Cherkash, Chick Bowen, Chimera101010, Chin Man2, Chipmunkdavis, Chmee2, ChrisO,

Image Sources, Licenses and Contributors

File:Moravian Slovak Costumes during Jizda Kralu.jpg *Source*: http://en.wikipedia.org/w/index.php?title=File:Moravian_Slovak_Costumes_during_Jizda_Kralu.jpg *License*: unknown *Contributors*: Jialiang Gao www.peace-on-earth.org

File:Český Šternberk2.jpg *Source*: http://en.wikipedia.org/w/index.php?title=File:Český_Šternberk2.jpg *License*: unknown *Contributors*: User:Marcus33

File:Husova bouda zima, lyzari.JPG *Source*: http://en.wikipedia.org/w/index.php?title=File:Husova_bouda_zima,_lyzari.JPG *License*: unknown *Contributors*: Petr Vilgus

File:Zámek Lednice.jpg *Source*: http://en.wikipedia.org/w/index.php?title=File:Zámek_Lednice.jpg *License*: unknown *Contributors*: Original uploader was Zp at cs.wikipedia

File:Czech crown jewels.jpg *Source*: http://en.wikipedia.org/w/index.php?title=File:Czech_crown_jewels.jpg *License*: unknown *Contributors*: User:Lamprus

File:Openstreetmap logo.svg *Source*: http://en.wikipedia.org/w/index.php?title=File:Openstreetmap_logo.svg *License*: unknown *Contributors*: OpenStreetMap

Image: Czechia - background map.png *Source*: http://en.wikipedia.org/w/index.php?title=File:Czechia_-_background_map.png *License*: unknown *Contributors*: Caroig

Image: Czechia - outline map.svg *Source*: http://en.wikipedia.org/w/index.php?title=File:Czechia_-_outline_map.svg *License*: unknown *Contributors*: Caroig

Printed by Books on Demand GmbH, Norderstedt / Germany